Chemistry of Milk and Milk Products

NIPA® GENX ELECTRONIC RESOURCES & SOLUTIONS P. LTD.
New Delhi-110 034

About the Author

Dr. Pinaki Ranjan Ray is currently employed with West Bengal University of Animal and Fishery Sciences in Kolkata as a professor and head of the Department of Dairy Chemistry. He has been conducting research and teaching in the field of dairy chemistry for over 23 years. In addition to mentoring numerous postgraduate students, he has over 60 research papers and technical articles published in reputable national and international journals. He received several honours and went to many national and international conferences in India and other countries. Dr. Ray is the author of one dairy chemistry textbook and has contributed chapters to several books published by domestic and foreign publishers. He is a lifelong member of SASNET and the Dairy Technology Society of India.

Advances in Processing, Preservation and Value Addition Technologies Volume 04

Chemistry of Milk and Milk Products

Pinaki Ranjan Ray
Professor
Department of Dairy Chemistry
Faculty of Dairy Technology
West Bengal University of Animal and Fishery Sciences
Mohanpur, Nadia 741252
West Bengal

NIPA® GENX ELECTRONIC RESOURCES & SOLUTIONS P. LTD.
New Delhi-110 034

NIPA® GENX ELECTRONIC
RESOURCES & SOLUTIONS P. LTD.

101,103, Vikas Surya Plaza, CU Block
L.S.C. Market, Pitam Pura, New Delhi-110 034
Ph : +91-11-43860225, Mob.: +91 9717133558, 9540816132
E-mail: newindiapublishingagency@gmail.com
Website: www.nipaersources.com

Print ISBN: 978-93-58871-64-7
ebook ISBN: 978-93-58875-22-5

Composed and Designed by NIPA®.

Dedication

with humility and reverence

this book is dedicated

at the lotus feet of my Late Grandmother

Smt. Rosy Ray

Preface

Dairy chemistry textbooks are widely available, with many of the authors being Western or Indian. What use does another text book serve, then? I have been putting off this question for a long time before starting this project. The chemistry of traditional Indian milk products is not often covered in Western textbooks, despite the fact that knowledge of this subject is crucial for Indian students. The chemistry of milk and milk products, including the chemistry of traditional milk products are not usually available under one cover. As a result, the students' study of dairy chemistry has seen them forced to rely on various text books. I've made an effort to include the fundamentals of milk chemistry, the chemistry of milk products, including the chemistry of traditional milk products from India, and the sophisticated analytical ideas needed to analyse them.

It provides a thorough synopsis of the chemistry of every type of milk product. Both the knowledge development during undergraduate study and the preparation for the postgraduate entrance exam will be greatly benefited by the discussions.

The end of each chapter includes a list of a few reference books. Although they don't cover every detail, these will be useful to students as a quick reference when studying dairy chemistry.

My reverend teachers, Prof. A.K. Bandyopadhyay and Prof. P.K. Ghatak, have shown me the way to write volumes of this kind, and for that I am grateful to them.

I would especially like to thank Dr. Soma Maji, an Assistant Professor in the dairy technology department at Centurion University of Technology and Management in Odisha, for her invaluable assistance in a number of ways while I worked on this book.

I am thankful to my postgraduate students, Payel Karmakar, Ankita Pandit, Chandrakanta Sen, Trisha Roy, Pratigya Pradhan, and Alisha Chhetri, for their assistance in preparing this manuscript.

I humbly ask for your forgiveness because I refuse to take on the nearly impossible task of remembering the names of every other colleague who has assisted me, either directly or indirectly, in publishing this book.

I am obliged to NIPA Genx Electronic Resources and Solutions P.Ltd, New Delhi, for providing me the opportunity to publish this book volume. I also want to thank NIPA's very talented and productive publishing team.

Finally, but just as importantly, I would like to thank my mother, Smt. Gayatri Roy, from the bottom of my heart. She has always inspired me to do work of this nature. I owe my wife Manjari Goswami and son Snigdho Ray as well for their unwavering support, cooperation, and encouragement.

Any suggestions for improvement of the book are most welcome.

Pinaki Ranjan Ray

Contents

Preface ... *vii*

1. **Chemical Nature of Milk ... 1**

Introduction ... 1

Definition of Milk ... 1

Gross Composition of Milk ... 1

FSSR (2011) Standard for Different Classes of Milk ... 2

Milk Secretion ... 2

Properties of Milk ... 3

Constituents of Milk ... 4

Other Constituents ... 7

Colostrum ... 9

Factors Affecting Milk Composition ... 10

Mastitic Milk ... 13

Effect of Mastitis on Milk Constituents ... 14

Changes in Lactose and Chloride ... 14

Changes in Milk pH ... 14

Enzymes ... 14

Miscellaneous Constituents ... 15

Heat or Oestrum ... 15

Gestation ... 15

References ... *15*

2. **Milk Protein ... 17**

Distribution of Protein in Milk ... 17

Fractionation of Milk Proteins ... 17

Non-Protein Nitrogen(Npn) of Milk ... 18

Fractionation of Milk Protein ... 18

Nomenclature of Milk Proteins ... 19

Minor Milk Protein ... 21

Whey Protein ... 27

Purification method 29
Casein 30
Amino Acid Composition 31
Characteristics of Different Types of Casein 32
References *35*

3. Milk Carbohydrate 37
Status of Lactose 37
Physico-Chemical Properties of Lactose 38
Proof of Lactose Structure 38
Physical Forms of Lactose 40
Mutarotation 42
Crystallization of Lactose 44
Reduction 50
References *50*

4. Milk Lipid 51
General Composition 51
Fatty Acid Profile of Milk Fat 54
Cholesterol 55
Phospholipids 55
Crystalline Morphology of Edible Fat 56
Lipid Oxidation 56
Rancidity 59
Type of Rancidity 60
References *61*

5. Milk Salt 63
Trace Elements 64
Distribution of Salts Between Casein Micelles and Serum 65
Physical Equilibria Among Salts 65
Salt Solution in Milk 65
Salt Balance in Milk 66
Salt Balance in Milk 66

Salt Balance Theory of Sommer and Hart 67
Factors of Salt Equilibria in Milk 68
References *69*

6. **Milk Enzyme 71**
There are Two Types of Enzymes in Milk 71
General Characteristics of Milk Enzymes 72
Common Milk Enzymes and their Role in Dairy Industry 72
References *78*

7. **Chemistry of Cream and Butter 79**
Chemistry of Creaming 79
Theory of Creaming 79
Cream Ripening 80
Physical Ripening 80
Ripening Process 82
Butter Churning 82
Factors of Churning 83
Biochemical Ripening 83
Nutrients in Cream 84
Flavor and Aroma of Cream 84
Composition and Standard of Butter 85
References *86*

8. **Chemistry of Cheese 87**
Nutritional Fact 87
Classification 87
Rennet 88
Cloned rennet 89
Chemistry of Rennet Action 89
Syneresis 90
Factors of Rennet Coagulation 91
Changes in the Biochemical Characteristics During Cheese Ripening 92
Accelerated Cheese Ripening 95

Acid Curd And Rennet Curd 95
FSSA (2006) Standard Of Various Cheeses 96
References *96*

9. Chemistry of Ice Cream 97
Standard of Ice Cream 97
Role of Different Constituents in the Ice Cream Mix 98
Physico-Chemical Effects of Different Mix Processing Operations 102
References *103*

10. Chemistry of Concentrated Milk 105
Sweetened Condensed Milk 105
Evaporated Milk 106
Nutritional Status of Concentrated Milk 107
Effect of Concentration on Milk Properties 107
Other Changes During Evaporation 108
Heat Stability of Concentrated Milk 108
Age Thickening and Gelation of Concentrated Milk 109
Lactose Crystallization in Sweetened Condensed Milk 110
References *111*

11. Chemistry of Dried Milk 113
FSSA (2006) Standard of Milk Powder 113
Physico-Chemical Changes During Manufacture of Dried Milk 114
Chemical Changes in Dried Milk During Storage 114
References *117*

12. Chemistry of Ghee 119
Composition 119
Role of Different Chemical Constituents 120
Chemical Constants of Ghee 123
References *124*

13. Chemistry of Traditional Indian Dairy Products 125
Classification of Traditional Indian Dairy Products 125
Chhana 126
Sandesh 128

Rasgulla 129
Paneer 129
Chemistry of Coagulation Process in Paneer 130
Factors Affecting Paneer Quality 131
Khoa 131
Chemistry of Khoa 132
Traditional Indian Fermented Dairy Products 133
Mishti Doi 135
Chakka and Shrikhand 136
Lassi 137
References *138*

14. Chemistry of Fermented Milk 139
Different Types of Fermented Milk Products 139
Changes in Nutrients Resulting from Fermentation 139
Impact of Fermented Milk on Nutrient Bioavailability and Digestibility 140
Flavor of Fermented Milk 140
Yogurt 141
Biochemical Changes in Yogurt 141
Beneficial Effects of Yogurt 142
References *143*

15. Milk Adulteration 145
Typical Adulterants in Milk 145
Qualitative Detection Method 146
Synthetic Milk 150
Negative Impacts of Synthetic Milk 151
References *151*

16. Advanced Techniques of Milk Analysis 153
Electrophoresis 153
Poly Acrylamide Gel Electrophoresis (PAGE) 153
Principle 154
Starch Gel Electrophoresis 158

Preparation of Casein for Starch Gel Electrophoresis 161
Paper Strip Electrophoresis 162
Technique 163
Chromatography 165
Column Chromatography 168
Paper Chromatography 170
Thin Layer Chromatography 172
Separation of Milk Fat Components by TLC 175
Composition of solvent A 176
Composition of solvent B 176
Affinity Chromatography 177
Gas Liquid Chromatography (GLC) 179
High Performance (Pressure) Liquid Chromatography (HPLC) 182
Ion Exchange Chromatography 188
The Separation of Amino Acids by Ion Exchange Chromatography ... 192
Method 192
Gel Filtration 193
Epoxy group 194
Spectrophotometry 198
Types of Spectroscopy 200
Principle of Spectrophotometer 200
Experiment 201
High Speed Centrifuge and Ultra-Centrifuge 202
Flame Photometry 207
Determination of Sodium and Potassium 208
Membrane Separation Process 209
Fractionation of Milk Protein by Ultrafiltration 210
Appendix 212
Spectrophotometer 213

17. Physical Properties of Milk and Milk Products 215
Physical Appearance, Color and Optical Properties 215
Flavor of Milk 215
Acidity and pH 221

Osmotic Pressure..222
Water Activity..223
Refractive Index..224
Specific Conductance..224
Ionic Strength..225
Thermal Conductivity..225
Thermal Diffusivity..226
References..226

18. Vitamins in Milk and Milk Products..227
Vitamins in Milk..227
Fat Soluble Vitamins..228
Vitamin D (Calciferol)..229
Vitamin E (Tocopherol)..230
Water Soluble Vitamins..233
Thiamine..233
Riboflavin..235
Folic Acid..236
Pyridoxine..237
Ascorbic Acid (Vitamin C)..239
References..240

Index..241

1

Chemical Nature of Milk

Introduction

The mammary glands of female mammals secrete milk, a necessary food that is liquid and meets the newborn's nutritional needs. All mammals consider milk to be a complete food for their newborns. It is considered a necessary food for humans and can be consumed as a product or as liquid milk. It contains almost all essential nutrients except iron, which is very low in milk. In essence, milk is an emulsion of protein and fat in water, supplemented with carbohydrates, vitamins, and minerals.

Definition of Milk

- Milk is a complex biological fluid that differs from species to species in both composition and physical properties. It is the full lacteal secretion of a mammary gland that is obtained by milking mammals for a minimum of 72 h after calving or until the milk is free from colostrum.
- FSSAI defines milk as the normal mammary secretion derived from the complete milking of healthy milch animals. It should be free from colostrum.

From a chemical perspective, milk is a dynamically balanced mixture that comes in different forms such as emulsion, colloidal suspension, and true solution. It also contains varying amounts of proteins, fats, carbohydrates, salts, and water.

Gross Composition of Milk

Milk composition varies among different mammals. The average chemical composition of milk in various species is given in Table 1. Buffalo milk and sheep milk contain higher amounts of fat, while human milk comparatively has the highest lactose content. Human milk has lower protein content as compared to milk of other species.

Table 1: Gross chemical composition of milk of different species.

Species	Composition (%w/w)				
	Water	Fat	Protein	Lactose	Ash
Cow	87.6	3.9	3.2	4.6	0.7
Buffalo	84.2	6.6	3.9	5.2	0.8
Human	87.7	3.6	1.8	6.8	0.1
Goat	86.5	4.5	3.5	4.7	0.8
Sheep	82.5	7.2	4.6	4.8	0.9
Camel	86.5	3.1	4.0	5.6	0.8

Source: Outlines of Dairy Technology (2004), Jenness and Patton (1959).

FSSR (2011) Standard for Different Classes of Milk

The Food Safety and Standards Authority of India prescribes specific standards for various types of milk. Milk standards for fat and SNF are different for species of buffalo, cow, goat, sheep, and camel. The standards of different classes and designations of milk are given in Table 2.

Table 2: Standards of milk for the Indian market.

Type of milk	Fat (Min.)	SNF (Min.)
Cow	3.2	8.3
Buffalo	6.0	9.0
Goat and sheep	3.0–3.5 (as per difference in-state)	9.0
Camel	2.0	6.0
Standardized	4.5	8.5
Skim	Not more than 0.5	8.7
Toned	3.0	8.5
Double toned	1.5	9.0
Full cream	6.0	9.0

Milk Secretion

The udder of a cow is a special organ that yields milk for the young. The mammary gland has four quarters and is made of fatty, connective, and mammary tissue, which secretes milk. Lactation is the term used to describe the intricate biochemical processes that the mammary gland goes through to produce milk. Milk is secreted in the tiny epithelial cells lining called "alveoli," which have a structure that resembles a bunch of grapes. The production of milk is linked to epithelial cells. Some milk constituents are formed by transferring blood constituents, while other constituents are synthesized from specific blood constituents.

Milk derives from blood, though exact amounts can differ. The milk yield decreases in the later stages of lactation. As milk passes through the udder, certain substances such as glucose, fatty acids, amino acids, vitamins, hormones, and minerals are extracted at the cell membranes separating the milk from the blood. The oxidation of both glucose and fatty acids is necessary for the passage of these materials into the mammary gland and the subsequent synthesis of milk. This process provides the energy needed for both transport and synthesis.

Milk's osmotic pressure (6.6 atm.) is the same as blood's. The conversion of amino acids into proteins and fatty acids into neutral fat is a step in the formation of milk. Approximately three-quarters of the total osmotic pressure of milk is accounted for by the combined concentration of lactose and chloride.

Physically, milk is a three-phase system made up of an emulsion of finely dispersed fat droplets, a colloidal dispersion of casein particles with calcium phosphate, and a phase of soluble proteins, minerals, vitamins, and salts.

Properties of Milk

Chemical Properties

1. Freshly drawn milk has a pH value ranges between 6.5 and 6.7 and contains 0.12–0.14% titratable acidity expressed as lactic acid. The acidity of the freshly drawn milk mainly happens due to the presence of CO_2, citrate, and casein. Natural acidity in milk is considered important from the heat stability point of view.
2. Fresh milk acts as a complex buffer due to the presence of CO_2, proteins, phosphates, citrates, and other minor constituents. This property of milk is considered important for curdling and heat stability.

Physical Properties

1. *Taste and odor:* Lactose milk sugar, and milk fat are the sources of the typical flavor and aroma, which is mildly sweet and aromatic. The flavor and aroma of milk are negatively impacted when it is produced in dirty conditions or by animals that are nearing the end of their lactation. The feeding of specific weeds or fodder, onions, garlic, and so forth, along with udder infection, will result in an unusual flavor and taste in milk.
2. *Colour:* Milk appears white in general because fat globules and other colloidally dispersed materials reflect light. The presence of the pigment carotene causes the color to become more intensely golden yellow. Once more, golden yellow grains and green forages allow this to enter the blood. Buffalo milk is completely white because the highly

abundant beta-carotene is completely converted into colorless vitamin A. Skimmed milk has a greenish tint after milk fat is removed, which is caused by cytochrome or riboflavin.

3. *Specific gravity:* Milk is heavier than water. Cow milk has a specific gravity that ranges from 1.025 to 1.032. It is less at higher temperatures and vice versa. It varies with temperature. The specific gravity of each component in milk is minerals (salts) 4.12, water 1.0, fat 0.93, protein 1.346, lactose 1.666.
4. *Boiling point:* At normal atmospheric pressure and temperature, water boils at 100 °C. The boiling point is raised to 100.15 °C in the presence of dissolved milk components.
5. *Freezing point:* This is the temperature at which the solid phase can melt or liquefy, and the liquid phase can freeze or crystallize. Under normal atmospheric pressure, pure distilled water freezes at 0 °C. However, milk freezes at a slightly lower temperature than water because it contains soluble ingredients that raise the boiling point, such as lactose and soluble salts. The average freezing point of cow or buffalo milk is −0.545 °C, but it can range from −0.535 to −0.55 °C. The freezing point will rise by 0.006 °C with the addition of 1% milk water.
6. *Viscosity:* A substance's resistance to flow brought on by intramolecular attraction is referred to as its viscosity. Milk's viscosity ranges from 1.5 to 2 centipoises (mPa s) at room temperature. Because milk contains dissolved solids, its viscosity is always higher than that of water.

Constituents of Milk

Chemical composition determines the ultimate nutritive value of the milk for the consumer and also has a direct effect on the output of various dairy products to be manufactured. For commercial purposes, it is usual to divide the milk constituents into (1) fatty and (2) non fatty solids also known as solid not fat (SNF). Water serves as the medium in which two sets of constituents are carried out.

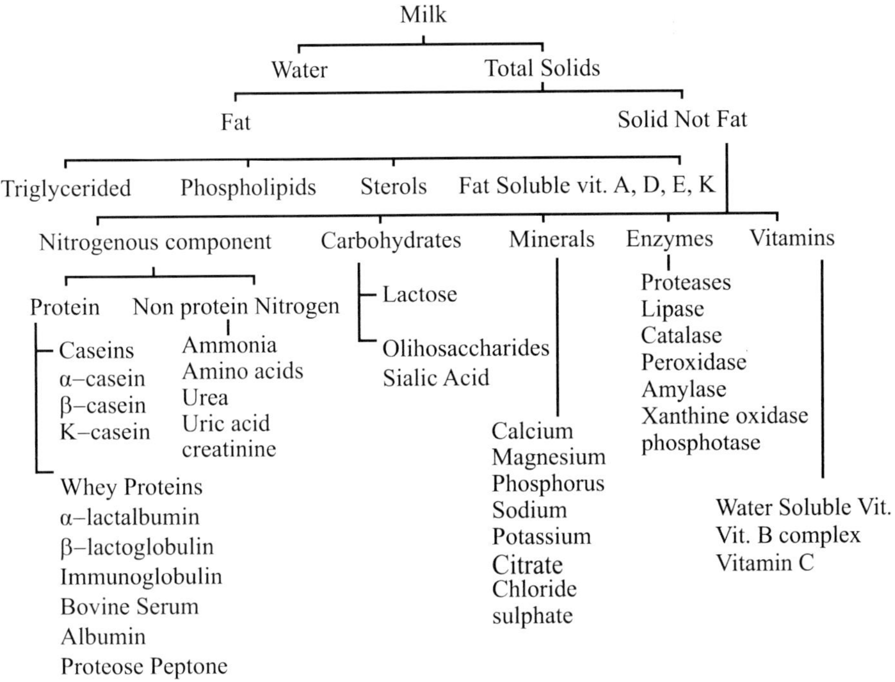

Fig. 1: Important constituents of milk.

Water: All milk constituents are either dissolved or suspended in water. Major part of the milk is constituted by water. Most of it is free and only a very small portion is in the bound form, being firmly bounded by milk proteins, phospholipids etc.

Total solids: Total solids constitutes lipids (fat) and solids not fat.

Milk fat (lipids): Small globules of milk fat, with an average size of 2–5 microns, are present in milk. It is an emulsion of the oil-in-water type. An adsorbed layer of material known as the milk fat globule membrane (MFGM) coats the globules. The phosphoprotein complex in this membrane keeps the fat emulsion stable. The fat globules are kept from clumping together and separating by the MFGM. However, agitation (at low temperature), heating, freezing, etc., can break the emulsion. Fatty acid glyceride-esters make up milk fat, which, when hydrolyzed, yields a mixture of fatty acids and glycerol. Milk contains both unsaturated and saturated fatty acids. Generally speaking, saturated fatty acids are stable. The compounds linked to fat include phospholipids, cholesterol, beta-carotene, and vitamins A, D, E, and K.

Phospholipids: Lecithin, Cephalin, and Sphingomyelin are the phospholipids present in milk fat. Lecithin, a component of MFGM, gives milk and other dairy products their distinctive flavor. It is extremely susceptible to oxidative changes, which give rise to metallic or oxidized flavors. In addition to their emulsifying properties, phospholipids aid in stabilizing the milk fat emulsion.

Cholesterol: Cholesterol is present in the true solution and forms a part of the MFGM complex with protein in the non-fat portion of milk.

Fat-soluble pigments: Carotene is the major fat-soluble pigment that is responsible for the yellow color of milk products. Carotene acts as an antioxidant and it also acts as a precursor of Vitamin A. One molecule of β-carotene gives two molecules of Vitamin A, whereas α-carotene gives one.

Fat-soluble vitamins: Milk is rich in fat-soluble vitamins i.e. A, D, E, and K.

Lactose, proteins, and minerals constitute the solid-not-fat (SNF) of milk.

Lactose: The main carbohydrate in milk is lactose. In the milk's serum portion, it appears as a true solution. Lactose crystallizes into tough, granular crystals. Sandiness in ice cream and condensed milk is caused by lactose. Fermented milk products are the result of the fermentation of lactose, which yields lactic acid and other organic acids. The production of lactic acid leads to souring, which is the cause of milk and milk product spoilage.

Milk proteins: Milk protein is constituted by casein, lactaglobulin, lactalbumin, milk serum albumin, and immunoglobulins.

The major milk protein is casein which constitutes more than 80% of the total proteins. It exists only in milk and is found in the form of a calcium caseinate phosphate complex. Casein in milk is present in colloidal state. It is precipitated by acid, rennet, alcohol, heat, and concentration. Casein is composed of α, β, κ, and γ fraction. α-casein constitutes 75% of casein and is responsible for the stabilization of casein micelle in milk. β and γ constitute 22 and 3% of casein, respectively. κ-casein is calcium-sensitive and responsible for rennet action in milk. α-Lactalbumin and β-lactoglobulin are the major whey protein components of milk. They exist in a colloidal state and are easily coagulated by heat. Milk serum albumin is the same as blood serum albumin of the blood. Immunoglobulins are mainly present in colostrum and give immunity to newborn calves.

Non-protein nitrogenous compounds: Ammonia, amino acids, protease-peptones, urea, and uric acid are non-protein nitrogenous compounds of milk.

Mineral matter or ash: Milk contains trace amounts of minerals and salts. They have a significant impact on the nutritional value and physicochemical properties of milk. Sodium, potassium, calcium, magnesium, phosphate, citrate, chloride, bicarbonate, and sulfate are the main salts. All other salts and minerals are considered as trace elements. After ashing, the mineral salts in milk are typically measured. Ash is basic, while milk is acidic. Mineral salts are found in two states: partially in true solution and partially in the colloidal state.

Other Constituents

Pigments: Water-soluble pigments are Riboflavin and xanthophylls. Riboflavin besides being a vitamin is a greenish-yellow pigment that gives characteristic color to whey.

Dissolved gases: Milk contains gases like O_2, CO_2, N_2, etc.

Water soluble vitamins: B complex and vitamin C.

Enzymes: Enzymes are biological catalysts. Milk contains enzymes like catalase, lipase, alkaline phosphatase, esterase, xanthine oxidase, lysozyme, ribonuclease, lactoperoxidase, etc.

Approximate concentration of the constituents in the normal cow milk are given in Table 3.

Table 3. Approximate concentration of the constituents in the normal cow milk

S.No.	Constituent/group of constituents	Weight/liters of milk
1.	Water	860-880 g
2.	Lipids in emulsion phase	
	Milk fat (a mixture of triglycerides)	30–50 g
	Phospholipids (Lecithin, Cephalin etc.)	0.3 g
	Cerebrosides	-
	Sterols	0.1 g
	Carotenoids	0.1–0.6 g
	Vitamin A	0.1–0.5 g
	Vitamin D	0.4 μg
	Vitamin E	1.0 mg
	Vitamin K	Trace
3.	Proteins in colloidal dispersion	
	Casein (α, β, γ fractions)	25 g
	β-lactoglobulin	3 g
	α-lactalbumin	0.7 g
	Albumin identical to blood serum albumin	0.3 g
	Euglobulin	0.3 g
	Pseudoglobulin	0.3 g

	Other albumins and globulins	1.3 g
	Fat globule protein	0.2 g
	Enzymes [catalase, peroxidase, xanthine oxidase, phosphatase (alkaline and acid), aldolase, Amylase (α, β), lipases and other esterases, proteases, carbonic anhydrase]	
4.	Dissolved materials	
	A. Carbohydrates	
	Lactose	40–45 g
	Glucose	50 mg
	Other organic sugars	Traces
	B. Inorganic and organic ions and salts	
	Calcium	1.25 g
	Magnesium	0.10 g
	Sodium	0.50 g
	Potassium	1.50 g
	Phosphates (as $PO4^{-}$)	2.10 g
	Citrates (as citric acid)	2.0 g
	Chlorides	1.00 g
	Bicarbonates	0.20 g
	Sulphate	0.10 g
	C. Water soluble vitamins	
	Thiamine	0.4 g
	Riboflavin	1.5 g
	Niacin	0.2–1.2 g
	Pyridoxine	0.7 g
	Pantothenic acid	3.0 g
	Biotin	50 µg
	Folic acid	1.0 g
	Choline	150 mg
	Vitamin B12	7.0 µg
	Inositol	180 mg
	Ascorbic acid	20 mg
	D. Nitrogenous materials not protein or vitamins (as N)	
	Ammonia	2–12 g
	Amino acid	3.5 mg
	Urea	100 mg
	Creatine	15 mg
	Uric acid	7 mg

	Uracil-4-carboxylic acid (orotic acid)	50–100 mg
	Hippuric acid	30–60 mg
	Indican	0.3–2.0 mg
	E. Gases	
	Carbon dioxide	100 mg
	Oxygen	7.5 mg
	Nitrogen	15.0 mg
	F. Miscellaneous	
	Esters of phosphoric acid not yet identified as phosphorus	0.1 g
5	Trace elements	
	Usually present Rb, Li, Ba, Sr, Mn, Al, Zn, B, Cu, Fe, Co, I Occasionally present or questionable Pb, Mo, Cr, Ag, Sn, Ti, V, F, Si	

Source: Outline of Dairy Technology (2004)

Colostrum

Colostrum is the first form of milk that is released by the mammary glands after giving birth. It's nutrient-dense and high in antibodies and antioxidants to build a new born calf's immune system. It greatly differs from milk produced later during lactation. Colostrum contains a higher amount of albumin, globulin, immunoglobulins, and minerals. A higher amount of Immunoglobulin imparts passive resistance to the newborn against pathogenic microorganisms. The colostrum of ruminants contains a trypsin inhibitor which protects the immunoglobulins from digestion. The globulin fraction in colostrum declines quite rapidly with successive milking.

Table 4: Comparative physical characteristics of colostrum and milk.

S. No.	Characteristics	Colostrum	Milk
1.	Taste	Slightly bitter	Sweet
2.	Odor	Abnormal	Normal
3.	Acidity	0.2–0.4	0.12–0.14
4.	Freezing point, °C	−0.606	−0.520 to −0.560
5.	Chloride, %	0.148–0.156	1.029–1.032
6.	Specific gravity	1.05–1.08	1.029–1.032
7.	Refractive index at 20 °C	More than milk	1.344–1.348
8.	Electrical conductivity	More than milk	0.005 mho
9.	Viscosity at 20 °C	More than milk	1.5 to 2.0 cp

Factors Affecting Milk Composition

The composition of milk varies greatly depending on the number of factors. The factors fall into two categories: (A) physiological or animal, which is determined by genetic makeup; and (B) environmental, which includes things like age, the number of lactations one has had in the past, pregnancy, nutritional status, etc. The variations in the mammal species that produce milk are the cause of the changes. Factors that affect the composition of milk are:

Physiological or Animal

Species: Species of animals largely affect the milk composition. Several factors influence the milk composition. It is very difficult to determine accurately the connection between a specific factor and its effect on composition. But milk specific to each species gives adequate nutrition to the newborns of that particular species. The milk composition of different species is given in Table1.

Breeds: The makeup of milk differs significantly between breeds within a given species. Among the main components of milk, fat content varies the most. This variation also affects other constituents, though not as much. A breed's milk with a higher fat content also has a higher SNF content, and vice versa. Table 5 lists the variations in composition caused by a species' breed.

Table 5: Variation of the composition of milk due to different breeds of cow.

Breeds of cow	Average composition (%)				
	Water	Fat	Protein	Lactose	Ash
Friesian	88.01	3.45	3.15	4.65	0.68
Sindhi	86.07	4.90	3.42	4.91	0.70
Gir	86.44	4,73	3.32	4.85	0.66
Tharparkar	86.58	4.55	3.36	4.83	0.68
Sahiwal	86.42	4.55	3.33	5.04	0.66

Source: Dairy Chemistry and Animal Nutrition by M.M.Rai(1964) Indian Dairy Products by Rangappa and Achaya, (1974).

Different quarters of udder: Udders of mammals have four distinct quadrants that are anatomically and physiologically distinct from one another. The first quarter of the udder has the highest milk fat content. The last quarter of the udder's milk has the lowest fat content. Right-hand quarter yields averaged marginally higher than those of the other quarters. When a milking machine is used to milk all four quarters at once, the fat percentage of the milk collected from the various quarters will be very similar. The milk fat content from each section of the udder is listed in Table 6.

Table 6: The fat content of milk from different parts of the udder (fat percent).

Animal	L. fore	R. fore	Average L. side	L. hind	R. hind	Average R. side
1	4.36	4.33	4.39	4.42	4.20	4.29
2	4.42	4.68	4.37	4.31	4.38	4.52
3	4.19	4.12	4.28	4.32	4.27	4.19

Source: Indian Dairy Products by Rangappa and Achaya, (1974).

Lactation period: The composition of milk undergoes significant changes as lactation progresses. The peak changes happen at the start and finish of the lactation cycle. The first secretion after parturition, called colostrum, is not the same as regular milk. It has less lactose and more casein, serum protein, total protein, and mineral salts. Although the amount of fat in milk varies greatly during the lactation phase, the overall trend in the percentage of fat is comparable to the trend in the solids-not-fat category. After delivery, the fat content is high, reaches a low point in the first or second month of lactation, and then gradually rises for the duration of lactation. Throughout lactation, the lactose content is essentially constant; however, it does start to decrease towards the end of the process. The total percentage of SNF is elevated at the initiation of lactation but declines steadily until 6–8 weeks postpartum. After this time, the increase in protein counters the decrease in lactose, so the SNF content tends to remain rather constant through the eighth month. The SNF content increases due to the circulating hormones of pregnancy.

The chloride content increases with advancement of lactation. Calcium decreases to a minimum concentration and then increases, whereas total phosphorus declines throughout. Natural acidity commences from a high level and decreases to the normal value during the first three months and remains fairly steady until the last month when it declines sharply.

Nutrition: The composition and yield of milk are significantly influenced by nutrition. To produce milk, the mammary gland takes up minerals, amino acids, fatty acids, glucose, acetate, and β-hydroxybutyrate from the blood. It is possible to modify an animal's diet to change the ingredients in milk. Protein can be changed by 0.6% and fat content by 3%, but diet manipulation does not affect lactose content other than excessive or insufficient feeding. Overfeeding beyond what is necessary to maintain the animals and maximize milk production generally results in no discernible changes to the composition of milk. The animal becomes overweight as a result of overfeeding. Conversely, underfeeding tends to lower milk yield and deplete the animal's stored energy.

Effects of diet on milk fat: Precursors, required by the mammary gland to synthesize fat are generated during fermentation of feed stuffs in the rumen. Diets that alter fermentation affect fat content. A low intake ratio of roughage to carbohydrate results in decreased acetate and butyrate, the major fat precursors. It increases propionate which decreases milk fat. The percentage of milk fat will increase linearly as the acetate-to-propionate ratio increases to 2.2, but not above this ratio. The minimum forage-to-concentrate ratio needed to maintain milk fat content is about 40:60. Forage length should be at least 0.64 cm as finely ground forages result in greater propionate production in the rumen, leading to reduced milk fat. Types of concentrate used in the ration also affect fat percentage. Processing grains used in concentrates by pelleting, grinding, etc. increases the digestion of starch in the rumen and the production of propionate which subsequently reduces milk fat. Cereal grains, used as concentrates, may be substituted with soluble carbohydrates (lactose, whey, and molasses) to alleviate the reduction in percentage fat caused by feeding low forage rations. These carbohydrates appear to promote the growth of rumen bacteria that produce acetate for fatty acid synthesis. Fat can be added to the diets of dairy animals at the rate of 5–6% of dietary dry matter to increase milk fat content. Saturated fats can only increase fat percentage whereas, unsaturated fats decrease fat percentage. Oilseeds also help in fiber digestion and acetate production thereby increasing fat percentage.

Influence of ration on milk protein: The milk protein fraction is affected the most by the energy content of the ration of the animals. A 0.45 reduction in SNF occur if energy is reduced by 40 percent. Amino acids are derived from microbes in the rumen and are largely required for milk protein synthesis. Thus, the milk protein content is influenced by factors that regulate microbial growth and the process of fermentation. Production of propionate in rumen increases the synthesis of milk protein by increasing the accessibility of certain amino acids like glutamate. Feeding of chopped forage (less than 0.64 cm) increases propionate, similarly increase in starch-digesting microorganisms promotes propionate production. Production of propionate in rumen increases milk protein synthesis by increasing the availability of certain amino acids like glutamate. Feeding of chopped forage (<0.64 cm) increases propionate, similarly increases in starch digesting microorganisms promote the production of propionate. Starch is needed to maintain microbial digestion and subsequent microbial protein synthesis. Both are positively correlated with milk yield and milk protein percentage. The source and processing of carbohydrates influence fermentation in the rumen and therefore affect milk composition e.g., starch in barley and wheat degrades more rapidly than that in corn. If degradation is slow milk yield and percentage protein decreases. Rumen microorganisms cannot utilize lipids as an energy source for the growth and synthesis of microbial protein. It should be noted that although milk protein percentage decreases when dietary fat is added, the protein yield is unaffected because of the increase in milk production from the energy-enriched ratio. Thus, it is recommended that when fat is added to the ration to increase its energy content, the diet should be enriched with amino acids supplements or nondegradable dietary intake of protein, for the protein percentage to be maintained in milk. Feeding excess protein does not appreciably alter the milk protein content, but the non-protein nitrogen content and at the time fat content may be enhanced. When the animals are allowed fresh pasturage, the SNF increases. The percentage of lactose in milk has not changed.

Diseases: Mastitis of dairy animals adversely affects both the quality and quantity of milk. Due to mastitis 15% reduction in production takes place during the lactation period. It also alters the composition of milk. This infection also reduces the shelf-life of fresh milk, and results in a fat reduction (5–12%), casein (6–18%), SNF (up to 8%), lactose (5–20%), and total solids (3–12%). It also causes an increase in sodium and chloride levels and a decrease in potassium levels in mastitic milk.

Mastitic Milk

Uterine infections can have a significant impact on changes in milk composition because they can result from pathological or physiological changes in the secreting cells. When mastitis-causing organisms enter the udder, the metabolic products may prevent them from being able to synthesize the components of milk as normally. The composition of milk is significantly influenced by udder infection, and milk from mastitic udders exhibits direct blood passage evidenced by high concentrations of red blood cells, sodium and chloride, and blood enzymes.

Effect of Mastitis on Milk Constituents

Changes in Protein

Casein production is decreased by mastitis, and the percentage of casein in milk is one of the main indicators used to examine the chemical alterations brought on by mastitis. Mastitis milk's rennet-induced clotting is significantly impacted and takes longer to set up. Rennet frequently fails to clot the abnormal mastitic milk significantly. Another characteristic of mastitic milk that is common in cheese making is decreased curd tension, red blood cells, and blood enzymes.

Milk from mastitic quarters contains increased amounts of non-casein proteins, especially blood proteins. This is due to increased capillary permeability during inflammation. Mastitic milk has higher levels of serum albumin Immunoglobulin and protease-peptone and lower levels of lactalbumin and lactoglobulin. Normal milk has an albumin: globulin ratio between 3 and 5, and mastitic milk has much lower ratios and these low ratios can be attributed to increased immunoglobulin.

Changes in Lactose and Chloride

Because of the altered osmotic equilibrium caused by mastitis, the percentage of lactose in milk from udders with mastitis is decreased. Because of the changed permeability, sodium chloride rises to the surface of milk and increases its osmotic pressure. Consequently, there is a noticeable increase in the chloride content of mastitic milk. The chloride concentration rises proportionately to the decrease in lactose number when mastitis causes a decrease in milk secretion. This rise is typically 1.5–3.0 for normal milk, but it rises noticeably in mastitis milk. Therefore, a high enough concentration of chlorides may be present in mastitic milk to give it a salty taste.

Changes in Milk pH

In mastitis, increased permeability of the gland to blood components results in higher values in milk. This might be partially due to the increased movement of the bicarbonate ion in to the milk. Since the lactose production decreases and the alkaline salts from the blood enter the milk, it becomes more alkaline, showing pH above 7.0.

Enzymes

The catalase enzyme exhibits elevated activity when an udder becomes inflamed. There is a correlation between the percentage of oxygen and the number of leucocyte cells, and this enzyme liberates molecular oxygen from hydrogen peroxide. Mastitis milk also has higher levels of the enzyme amylase. Since *Streptococci* sp. can hydrolyze phenyl phosphate substrate at pH 4.1, milk infected with this organism was found to have elevated levels of the enzyme acid phosphatase. Three esterases are present in normal milk: A, B, and C. Of these, it has been observed that mastitic milk has significantly higher levels of A-esterase.

Miscellaneous Constituents

It has been reported that when the concentration of vitamin A in milk increases, so does the total amount of carotenoid. It has been shown that increased permeability between blood and milk could be the cause of the higher β-carotene levels in mastitic milk. In mastitic milk, thiamine, a water-soluble vitamin, was reduced by 10%, and riboflavin by 20%. It has also been demonstrated that mastitic milk has reducing qualities that normal milk does not. In addition to chlorides, it has been established that low levels of calcium, phosphorus, and potassium are present in mastitis-affected milk. Chloride levels rose in tandem with sodium. There have been reports of increased concentrations of neuraminic acid and nine-fold increases in glycogen in mastitic milk.

Heat or Oestrum

Although not constant, oestrum does have an impact on the composition of milk. The fat test typically varies a lot, and the yield also slightly decreases over time. Increased anxiety and excitability have been linked to these alterations, which have caused the cow to either hold back some of the milk or secrete less of it.

Gestation

By accelerating the end of lactation, gestation can indirectly alter the composition of milk, leading to noticeable alterations in its composition. During a similar stage of their lactation, open cows exhibit no increase or a very slight increase in their solids content.

References

De, Sukumar. Outlines of Dairy Technology. New Delhi: Oxford University Press, 2004.

Fox, P. E. and P. L. H. McSweeney. Dairy Chemistry and Biochemistry. Madras: Blackie Academic & Professional, Chapman & Hall, 1998.

FSSA. The Food Safety and Standard Act. New Delhi: Professional Book Publishers, 2006.

Jenness, R. and S. Patton. Principles of Dairy Chemistry. New York: John Wiley & Sons Inc., 1959.

Mathur, M. P., D. Datta Roy, and P. Dinakar. Text Book of Dairy Chemistry. New Delhi: Indian Council of Agricultural Research. Govt. of India, 1999.

Rai, M. M. Dairy Chemistry and Animal Nutrition. Agra: Ram Prasad & Sons, 1964.

Rangappa, K. S. and K. T. Acharya. Indian Dairy Products. Bombay: Asia Publishing House, 1974.

Webb, B. H., A. Johnson, and J. A. Alford. 1978. Fundamentals of Dairy Chemistry. West Port: The AVI Publ. Co. Inc.

2

Milk Protein

Bovine milk contains an average of 3.5% protein. The concentration of protein changes significantly during lactation. The first few days of lactation exhibit the main change in protein concentration. The main changes occur in whey protein fraction.

Distribution of Protein in Milk

80% casein and 20% whey protein (β lactoglobulin-10%, α-lactalbumin-5%, serum albumin, and immunoglobulin-5%).

Table 1: Average casein and whey protein of different milk.

Type of milk	Casein	Whey protein
Cow milk	2.8%	0.6%
Goat milk	2.5%	0.4%
Sheep milk	4.6%	0.9%
Human milk	0.4%	0.6%

Source: Fox and Mcsweeney (1998), Dairy Chemistry and Biochemistry, Blackie Academic and Professional, p. 148.

Fractionation of Milk Proteins

Fractionation of milk protein is mainly done by acidification at pH 4.6 around 30 °C.

Almost 80% of the bovine milk protein precipitates at pH4.6, and this fraction yielded by precipitation is called casein.

The remaining soluble part of the protein is referred to as whey protein/serum protein/non-casein protein.

Wide variation is observed in the casein-whey protein ratio in different types of milk(protein synthesis is independent of animal diet)

Significant Difference Between Casein and Whey Protein

1. Casein precipitates at pH 4.6 but does not precipitate whey protein at that pH. This property of milk protein is commercially exploited for the preparation of industrial casein and certain varieties of cheese.

2. Chymosin or other proteolytic enzymes produce specific changes in casein resulting in coagulation in the presence of calcium. Whey protein does not undergo this type of coagulation.
3. Casein is very stable to high temperatures. They can withstand heating up to 140 °C for 20 minutes but whey protein is relatively heat labile. They are completely denatured by heating at 90 °C for 10 min.
4. Casein is a phosphoprotein containing an average of 0.85% phosphorous while whey protein does not contain any phosphorous. The phosphate of casein is an important contributor to remarkably high heat stability and calcium-induced coagulation of rennet-altered casein.
5. Casein contains very low sulfur (0.8%) whereas whey protein is relatively rich in sulfur (1%). The principle casein contains only methionine but whey protein contains a significant amount of cystine, cysteine, and methionine. These sulfur-containing amino acids are responsible for many changes in milk on heating.
6. Casein is synthesized only in the mammary gland and no other sources in nature. Some of the whey proteins β lactoglobulin, and α-lactalbumin are synthesized in the mammary gland but bovine serum albumin and immunoglobulin are not synthesized in the mammary gland but derived from blood.
7. The whey protein is dispersed in solution and has a simple quaternary structure. Casein has a complex quaternary structure and exists in milk as large colloidal aggregates.

Non-Protein Nitrogen(NPN) of Milk

It constitutes 20–37% of nitrogen constituents. NPN in milk are urea nitrogen (5–10%), amino nitrogen (2.5–7%), creatinine (1–1.5%), creatin (2–2.5%), and uric acid (1–2%).

Fractionation of Milk Protein

The fractionation of milk protein was first developed by Rowland (1939). It determines the concentration, characteristics, and other behavior of milk protein. The Rowland scheme of fractionation was revised by Aschaffen Burg and Drewry in 1959.

Mellander identified casein as α, β, and γ using free boundary electrophoresis. This represents 60, 17.6, and 2.4% of total casein, respectively. Apart from the other fractions,it also contains whey protein, euglobulin, pseudoglobulin, component 3, α-lactalbumin, β-lactoglobulin, and blood serum albumin until 1939. The primary structure of bovine milk casein suggests that casein consists of four types of gene variants: α_{S1}casein, α_{S2}casein, γ casein, k casein. Others are derived from post transitional processing such as phosphorylation, glycosylation, or limited proteolysis. The major whey protein is α-lactalbumin, β-lactoglobulin, immunoglobulin, and serum albumin. The protease peptone was also formed from limited proteolysis of other casein. Further it was found that there are seven variants of β-lactoglobulin, two variants of α-lactalbumin, four variants of α_{S1}caseinandα_{S2} casein, and two variants of k casein.

Table 2: Average composition of casein fraction in milk.

Casein type	Weight (%)	Mole (%)	Casein/skim milk (g/L)	Casein/skim milk (mole/L)
α_{S1}	38.1	36.4	10.25	3.52
α_{S2}	10.2	9	2.74	1.25
β	35.7	33.6	9.6	2
κ	12.8	14.8	3.45	0.19

Source: Textbook of Dairy Chemistry (2005), ICAR, New Delhi, p. 33.

Nomenclature of Milk Proteins

1. ADSF first formulated a committee on protein nomenclature. The nomenclature of proteins was first outlined in the works done by Jenness et al.(1956), Brunner et al.(1960), Thompson et al.(1965), and Roseet et al.(1970) and has been accepted by almost all workers in the field of dairy science.
2. The nomenclature developed for bovine milk proteins such as α-lactalbumin, β-lactoglobulin, and k-casein has been employed for the proteins of other milk that are homologous to those in bovine milk.
3. Casein includes all the protein fractions that are considered to be casein and exclude all others. Solubility criteria at the isoelectric point present in caseinate micelle or content of ester-bound phosphates satisfactorily differentiate casein from all other portions. The casein of cow milk was found by Mellander (1939) to be resolvable in moving boundary electrophoresis at pH 8.6 into distinct peaks designated as α, β, ϒ in order of their decreasing mobility. However, each peak in such a pattern does not represent an individual component. The fractionation process was later separated with mobilities corresponding to the peaks obtained with the casein thus named α, β, ϒ. Subsequent fractionation and improved electrophoretic resolution indicated that α-casein consists of components sensitive to precipitation by calcium (α_s) and insensitive to such precipitation (k-casein). α_s casein is not entirely homogeneous. Its

main fraction has been designated as α_{s1} casein. Its other components have been isolated and fractionated consisting of at least 3 components named as α_{s2}, α_{s3}, α_{s4}.

4. Genetic variants of α_{s1}, β, ϒ, and k-casein have been discovered later and have been designated as α_{s} casein A, k-casein A, k-casein B. α_{s1} casein has five variants, α_{s2} has four genetic variants, β casein has seven variants, α-lactalbumin has two variants and β-lactoglobulin has seven variants.
5. Non-casein fraction or whey protein nomenclature was early resolved in lactoglobulin fractions. Soluble salt concentrations such as 50% saturated ammonium sulfate and lactalbumin fraction are insoluble in this environment. That indicates the presence of twoprinciple kinds of particle of different size (α and β). Lactalbumin was obtained by ultra-centrifugal analysis by Pedersen (1956). Later, when the protein such as α-lactalbumin, and β-lactoglobulin were crystallized from lactalbumin fraction the Greek letters (α-lactalbumin, β-lactoglobulin) were incorporated into the names. β protein was nomenclature as lactoglobulin because of its insolubility near the iso-electric point in the absence of salt and the α protein as an albumin due to the source in the lactalbumin fraction of whey. The genetic polymers of β-lactoglobulin are designated by the usual ABC system. 2 α-lactalbumin, A and B have also been found.
6. Protein identified in bovine blood serum albumin has been isolated from cow milk. The milk of every species examined electrophoretically contains the component with the mobility of the serum albumin. This protein, when isolated, from milk will be indicated by the term bovine serum albumin.
7. Milk contains antibody protein produced by the lactating female in response to the antigen to which she has been exposed. This antibody is synthesized elsewhere in the body than in the mammary gland and transferred to milk via the bloodstream. They may be synthesized also by the plasma cells localized in the mammary tissue upon fractionation of whey protein they are found almost in the globulin fraction insoluble in a 50% saturated solution of ammonium sulfate. The term ϒ globulin has been often applied to antibody proteins of blood serum as they migrate in the slowest-moving zone of electrophoresis. The generic name used is ϒ globulin.

8. The nomenclature of immunoglobulin to antibody fraction of milk was promulgated the name I_g. They are heterogenic group. The class of I_g recognized by the WHO are nomenclature as I_g (A, G, M), respectively designated as I_gG or ϒ-G, I_g M or ϒ -M, I_g A or ϒ-A.
9. Two kinds of iron containing protein have been detected in milk. One of this is identical to circulating transferrin in blood. It has been found in bovine milk and probably occurs in milk of other species. The other very different protein occurs in such secretions as saliva, tears, and semen. It is called red-protein lactotransferrin or lactoferrin.

Minor Milk Protein

Lactoferrin

Lactoferrin is an iron-binding glycoprotein originally isolated from milk. It was first identified in 1939 in bovine milk by Sorensen and Solensen. It is the glycoprotein of the lactoferrin family.

Occurrences and Level

Lactoferrin is predominantly found in the products of exocrine glands located in the gateway of the digestive, respiratory, and reproductive systems. The level of lactoferrin is significantly lower in bovine milk than in human milk. Approximately 30% of iron in human milk is bound to lactoferrin. Lactoferrin levels in human milk does not depend on the body's iron status. It is dependent on the general statement of maternal nourishment.

Table 3: Occurrence of lactoferrin in biological fluid.

Biological fluid	Amount
Colostrum breast milk	7 mg/mL
Matured breast milk	1–2 mg/mL
Cow's colostrum	1.5 mg/mL
Cow's milk	20–200 µg/mL
Tear fluid	2.2 mg/mL
Seminal plasma	0.4–1.9 mg/mL
Synovial fluid	10–18 µg/mL
Saliva	7–10 µg/mL

Refs: Korhonen (1977), Arnold et al. (1979), Levay and Vilijoen (1995).

It was revealed that Lactoferrin concentration changes during the mammary cycle in the cow. The concentration of lactoferrin and mRNA increase during development of the mammary gland and colostrum formation and involution of the gland.

During lactation the levels of lactoferrin decrease as opposed to the increasing level of casein. The latter suggests that casein has primarily nutritional function in calf whereas lactoferrin may have growth factors like activity and protection function for the newborn in the critical phase of their path.

Structure

Lactoferrin is a single chain polypeptide of about 80kDalton molecular weight containing 1-4 glycan depending on the species. Bovine and human lactoferrin consist of 689 and 691 amino acids respectively. Lactoferrin comprises two homologous lobes containing N and C lobes referring to the N-terminal and C terminal of the molecule. Each lobe further consists of twosub-lobes or domains which form a cliff where the ferric iron keeps tightly bound in synergistic co-operation with the bicarbonate anion.

Physio-Chemical Properties of Lactoferrin

Lactoferrin has a very high iso-electric point. The theoretical pI values calculated for bovine and human are 9.4 and 9.5. Depending on the methods used values around eight have been reported for bovine lactoferrin whereas a wide range of iso-electric pH of 5.5–10 have been reported for human lactoferrin and these variations may be due to variation in the arginine-rich N-terminal of the molecule due to separation condition.

Biological Functions of Lactoferrin

Different functions have been attributed to lactoferrin. It is considered a key component in the host, 1st line of defense as it can respond to a variety of physiological and environmental changes. The following immune-modulatoryfunctions are attributed to lactoferrin:

- Antioxidant activity.
- Antimicrobial activity against a broad spectrum of bacteria, fungi and yeast, viruses, and parasites.
- Anti-inflammatory activity.
- Anti-carcinogenic activity.
- Enzymatic function.
- Anti-parasitic activity.
- Immuno modulatory approach.

Bio Active Peptides from Lactoferrin

Enzymatic treatment of bovine lactoferrin with pepsin produces low molecular weight peptide with antibacterial properties against a large number of

Gram-positive and -negative bacteria including *E. coli, Salmonella enteritidis, Protease vulgaris*, and *Streptococcusbovis*. Belani etal. (1992) identified a region of amino acid at N-terminal that retains biological activity from full molecule called Lactoferricin B shows greater antimicrobial activity of peptides.

Immunoglobulin

Immunoglobulin is an antibody or Υ globulin.

Classification

There are five types of immunoglobulin (Ig). IgA, IgG, IgE, IgD, and IgM. Among these milks are IgG, IgA, and IgM. IgG has two sub-groups: IgG_1 and IgG_2.

Amount of Ig in milk:

Mature milk contains 0.6–1 g/L.

Colostrum contains 100 g/L.

Ig decreases rapidly as the lactation period advances.

Physio-chemical properties:

1. Ig constitutes approximately 75% antibody in an adult.
2. IgG is transferred from mother to child "utero" via cord blood and by breast feeding and serves as a child's first line of immune defense referred to as passive immunity.
3. IgA is secreted in breast milk and ultimately transferred to the digestive tract in the newborn infant, providing better immunity than a bottle-fed child.
4. Colostrum contains significantly greater concentration of Ig than matured milk. Ig reaches maximum concentration in the first 24–48 h post parturition and decline in a time dependent manner.
5. The whey fraction of milk appears to contain significant amount of Ig approximately 10–15%.
6. In vitro study demonstrated bovine milk derived IgG suppresses human lymphocyte proliferation at levels as low as 0.3 mg/mL IgG.

Structure of Ig

Ig are very complex proteins. IgG consists of two very long and heavy and two short and light polypeptide chains linked by dipeptides. IgA consists of two such units i.e. each chain is linked together by a secretory component and a junction component. IgM consists of five linked four-chain units. The heavy and light chains are specific to each type of Ig.

Secretion of Ig in Milk

IgG of bovine colostrum is derived exclusively from blood plasma. It is presumed that cellular uptake involves the binding of IgG molecules through Fe fragments to receptors situated in the basal membrane just before parturition.

Before parturition there is a sharp increase in the number of receptors showing high affinity for IgG was observed.

$IgG_1$is selectively transported into bovine colostrum. The mechanism mostly involves vesicular transport followed by exocytosis at apical membranes.

IgA in colostrum is derived partly from intra-mammary synthesis and partly by accumulation in the gland after being transported in the blood from other sites of synthesis.

IgA molecules are transported into the secretory cells across the basal membrane using a large membrane-bound form of the secretory component which acts as a recognition site.

The IgA complex is formed by endocytosis and transportation to the apical membrane of the secretory cell where cleavage of a complex portion takes place and the matured IgA is secreted by exocytosis.

Lipoprotein (LP)

Lipoprotein (LP) contains both protein and lipids bound to proteins which allows fat to move through the water inside and outside the cell. The protein usually emulsify the lipid molecule. Many enzymes, structural proteins, antigens, transporters, and toxins are lipo proteins. The example includes plasma LP particles and HDL, LDL, VLDL. These LP enable fat to be passed in the blood stream. The mitochondrial transmembrane proteins along with the chloroplast and the bacterial LP are the examples of LP.

Classification of LP

On the basis of density LP are classified into:

- Chylomicrons
- Very low density lipo proteins (VLDL)
- Intermediate density lipo protein (IDL)
- Low density lipo proteins (LDL)
- High density lipo proteins (HDL)

Chylomicrons

It carries triglycerides from the intestine to the liver, to the skeletal muscle and to the adipose tissue.

VLDL

Carries newly synthesized triglycerides from the liver to the adipose tissue.

IDL

It is an intermediate between VLDL and LDL. They are not detectable in blood.

LDL

It carries wide range of fat molecules. It carries 3000–6000 fats (phospholipids, triglycerides, cholesterol, etc. are carried by LDL). LDL are sometimes referred as bad LP because its concentration are related to atherosclerosis formation.

LDL is classified in to two types

- Large buoyant LDL (ldLDL)
- Small density particle LDL (sdLDL)

HDL

They collect fat molecules such as phospholipid, cholesterol, triglyceride from the body cell or tissue and take it back to the liver. HDL is considered as a good LP because higher concentration of LP correlates with low rate of atherosclerosis progression or regression.

Physical Properties of LP

Chylomicron

Density	<0.9 g/mL
Diameter	100–1000 nm
Protein	<2%
Cholesterol	8%
Phospholipid	7%
Triacylglycerol and cholesterol ester	84%

VLDL

Density	0.95–1.006 g/mL
Diameter	30–80 nm
Protein	10%
Cholesterol	22%
Phospholipid	18%
Triacylglycerol and cholesterol ester	50%

IDL

Density	1.006–1.019 g/mL
Diameter	25–50 nm
Protein	18%
Cholesterol	29%
Phospholipid	22%
Triacylglycerol and cholesterol ester	1%

LDL

Density	1.019–1.063 g/mL
Diameter	18–28 nm
Protein	25%
Cholesterol	50%
Phospholipid	21%
Triacylglycerol and cholesterol ester	8%

HDL

Density	>1.063 g/mL
Diameter	5–15 nm
Protein	33%
Cholesterol	13%
Phospholipid	29%
Triacylglycerol and cholesterol ester	4%

According to serum protein in electrophoresis:

- α-LP
- β-LP

Milk Fat Globule Membrane (MFGM)

Table 4: Gross composition of MFGM.

Component	**Fat globule (mg/100 g)**	**Fat globule area (mg/m^2)**	**%(w/w) of total membrane**
Protein	900	4.5	41
Phospholipid	600	3	27
Cerebrosides	80	0.4	3
Cholesterol	40	0.2	2
Neutral glycerides	300	1.5	14
Water	280	1.4	13
Total	2200	11	100

Source: Fox and Mcsweeney (1998), Dairy Chemistry and Biochemistry, Blackie Academic and Professional.

Mass spectrometry and other analytical methods characterize protein and lipid isolated from bovine MFGM. The major MFGM protein composition consists of xanthine oxidase, butyrophillin, acidophillin. The minor MFGM consists of polymeric Ig receptor protein, apo LP-E, apo LP-A-1, heat shocked cognate-P, clusterin, lactoperoxidase, Ig with heavy chain, peptidyl prolyn isomerase-K, actin, fatty acid binding protein and mucin (total MFGM consists of 1% of total protein.

Function

Xanthine oxidase: requires Fe, Mo, FAD as cofactor is capable of oxidizing lipid via the production of superoxide dimutase. 20% of MFGM protein is xanthine oxidase.

Butyrophillin: it is another important MFGM protein. It is named as such because of its high affinity for milk lipids. It is highly hydrophobic and difficult to solubilize glycol protein. Butyrophillin consists of 526 amino acids. It binds phospholipids and contains covalently bound fatty acids. It is located only at the apical cell surface of mammary epithelial cells. It has a role in the membrane envelopment of fat globules.

Whey Protein

α Lactalbumin and β Lactoglobulin

About 20% of thc total protein of bovine milk belongs to a group of proteins generally referred to as whey protein/serum protein/non-casein nitrogen. These are αlactalbumin, βlactoglobulin, blood serum albumin, and immunoglobulin.

α Lactalbumin (α LA)

Occurrence: α LA represents about 20% of the proteins of bovine whey (3.5% of the total milk protein). It is the principle protein in human milk. It is the principle protein with a molecular mass of 14 kD.

Amino Acid Composition of α LA

α LA is relatively rich in tryptophan.

It is also rich in sulfur (1.9%) which is present in cysteine, cystine, and methionine.

The principle α LA contains more phosphorous and carbohydrate.

Table 5: Amino acid composition of α LA.

Amino acid	β Lactoglobulin	α Lactalbumin
Asp	11	9
Asn	5	12
Thr	8	7
Ser	7	7
Ser P	0	0
Glu	16	8
Gln	9	5
Pro	8	2
Gly	3	6
Ala	14	3
½ Cys	5	8
Val	10	6
Met	4	1
Ile	10	8
Leu	22	13
Tyr	4	4
Phe	4	4
Trp	2	4
Lys	15	12
His	2	3
Arg	3	1
Pyroglu	0	0
Total residue	162	123
Molecular weight	18362	14174

Source: Fox and Mcsweeney (1998), Dairy Chemistry and Biochemistry, Blackie Academic and Professional.

Genetic Variants

Mainly two genetic variants of α LA are observed in bovine milk. These are called α LA A and α LA B.

α LA A does not contain any argenine whereas 1 argenine residue of α LA B is replaced by glutamic acid.

Structure of α LA

α LA has four structure: primary, secondary, tertiary, and quaternary.

Primary structure: The primary structure of α LA is similar to the structure of lysozyme. Out of the total 123 residues in α LA, 54 residues are identical to corresponding residue in lysozyme and further 23 residues are structurally similar.

Secondary structure: α LA is a compact globular protein that exists in a solution as a prolate ellipsoid with dimensions of 2.5*3.7*3.2 nm.

It consists of 26% α-helix structure, 14% β structure, and 60% unordered structure.

Tertiary structure: The tertiary structure of α LA is similar to the lysozyme structure. It is very hard to crystallize bovine α LA by X-ray crystallography.

Quaternary structure: Quaternary structure of α LA and its associates under various environmental conditions is not yet properly studied

Characterization of α LA is done by: electrophoresis, Brad Ford Assay, and UV spectrophotometry.

Purification method

1. Ion exchange chromatography
2. Adsorption chromatography
3. FPL chromatography, etc.

Biological function α LA is the main component of lactose synthetase which catalyzes the final step of biosynthesis of lactose. Lactose synthetase is composed of twoproteins named protein A and protein B. Protein A is UDP galactosyl transferase while protein B is α LA. When α LA is absence, protein A acts as nonspecific galactosyl transferase which transfers galactose from UDP galactose to different acceptors but in presence of α LA/protein B it becomes highly specific and transfers galactose only to glucose to form lactose. The synthesis of lactose is controlled directly by α LA.

β-Lactoglobulin (β Lg)

Occurrence

β Lg is a major protein in bovine milk which represents about 50% of the total whey protein and 12% of the total protein present in milk. β Lg is a principle whey protein in bovine, caprine and buffalo milk.

Heterogeneity

Bovine β Lg has four genetic variants which are designated as A, B, C, and D.

(Amino acid composition of β Lg)

β Lg is rich in sulfur-containing amino acids which gives it a high biological value of 110. β Lg contains two molecules of cysteine and one molecule of cysteine. The cysteine is specifically important because it reacts with the disulfide of k-casein during heat denaturation and significantly affects rennet coagulation and heat stability properties of milk.

Structure of β Lg

Primary Structure

β Lg consists of 162 amino acid residues.

Secondary Structure

β Lg is a highly structured protein. Optical rotary dispersion and circular dichroism measurement show that in the pH range of 2–6, β Lg consists of 10–15% α helix, 46% β sheet, and 47% unordered structure.

Tertiary Structure

The tertiary structure of β Lg is normally studied by using X ray crystallography. β Lg have very compact structure in which the β sheath occurs in a β barrel type of structure or sheet. Each monomer exists almost as a sphere with a diameter of about 3.6 nm.

Quaternary Structure

β Lg shows interesting association characteristics. Below pH 3.5 β Lg dissociates to monomers of 18 kD between pH 5.5 and 7.5 all β Lg form dimmers of 36 kD molecular mass. This dimer consists of A and B variant between 3.5 and 5.2 especially at pH 4.6 bovine β Lg forms octamers of molecular mass 144 kD. β Lg A associates more strongly than β Lg B because it contains an additional aspartic acid instead of glycine which forms an additional hydrogen bond.

Biological Function

- β Lg can bind retinol in the hydrophobic pocket and protect it from oxidation.
- It helps in the transportation of retinol through the stomach to the small intestine where the retinol is transferred to retinol-binding protein which has a similar structure to β Lg.
- β Lg also binds free fatty acid and thus it stimulates lipolysis.

Casein

The primary protein in milk is casein, which makes up 80% of the total protein content. Whey or serum proteins make up the remaining 20%. The fundamental ingredient in regular cheese is casein. Rennet enzymes cause casein to precipitate during the cheese-making process, forming a coagulum that includes casein, whey proteins, fat, lactose, and milk minerals.

Coagulation by rennet or precipitation by acid are the two common processes used to make commercial casein from skim milk. By multistage washing in water, as much fat, whey proteins, lactose, and minerals as possible must be eliminated because they lower the casein's quality and preservation quality. Properly produced and dried casein is mostly used in the food and chemical industries and has a good shelf life.

Physico-chemical Properties of Casein

Chemical composition of casein

The principle chemical or physicochemical properties of principle milk protein are given in the following table:

Property	**α_{s1}-B8P**	**α_{s2}-A11P**	**β-A^2 5P**	**k-B1P**
Molecular weight	23,614	25,230	23,983	19,023
Residues/molecule				
Amino acids	199	207	209	169
Proline	17	10	35	20
Cysteine	0	2	0	2
Intramolecular disulphide bond	0	0	0	0
Phosphate	8	11	5	1
Carbohydrate	0	0	0	-
Hydrophobicity(kJ/residue)	4.9	4.7	5.6	5.1
Charge				
Molecule%/residue	34	36	23	21
Net charge	–0.10	–0.07	–0.006	–0.02
Distribution	Uneven	Uneven	Very uneven	Very uneven
A_{280}	10.1	14.0	4.5	10.5

Source: Fox and McSweeney (1998), Dairy Chemistry and Biochemistry, Blackie Academic and Professional.

Amino Acid Composition

The major 4 features of the amino acid profiles of casein are:

- All caseins contain a high amount (35–45%) of polar amino acids (valine, leucine, isoleucine, phenylalanine, tyrosine, and proline), which would normally make them poorly soluble in an aqueous solution. However, the influence of polar amino acids is lessened in k-casein because of its high carbohydrate content, low level of sulfur-containing amino acids, and high phosphate group content.

- There is a very high proline content in all casein. Proline residues/moles of α_{s1}, α_{s2}, β, and k casein are 17, 10, 35, and 20 respectively. The casein's α-helix or β-sheet structure is very low due to the high levels of proline content. As such, the caseins are easily subjected to proteolysis without the need for previous denaturation
- The biological value of caseins is limited due to their deficiency in sulfur-containing amino acids (80 for casein). While α_{s2} and k casein has two cysteine residues/moles that typically exist as intramolecular disulfides, α_{s1} and β casein do not contain any cysteine at all. Whey protein is the main milk protein that contains sulphydryl. 1-SH group is present in β-lactoglobulin. These -SH groups are typically passive and unresponsive. The -SH group of β-lactoglobulin becomes exposed and reactive during heat denaturation at temperatures above 75°C, and it exchanges sulphydryl disulfide with k casein. These phenomena have a major impact on heat stability and rennet coagulation, two of the physicochemical properties of milk that are crucial to technology
- Lysine and other essential amino acids are abundant in caseins, particularly α_{s2} casein. Casein is an excellent nutritional supplement for lysine-deficient cereal proteins. Owing to the high lysine content, when heated in the presence of reducing sugar, casein and its product undergo extensive Maillard browning. The main cationic residue in casein is lysine, with smaller amounts of histidine and arginine.

Characteristics of Different Types of Casein

α_{s1}-casein

1. The polypeptide chain of α_{s1} - casein consists of two predominantly hydrophobic regions (residues 1–44 and 90–199) and a highly charged polar zone
2. Prolines are sporadically found in the hydrophobic segments, and all but one of the phosphate groups are located in the 45 and 89 residues segment. As a result, this protein can be seen as a flexible, relatively loose polypeptide chain
3. Self-association of α_{s1} – casein largely depends on pH, ionic strength, and type of ion in the medium, but is largely temperature independent

β-casein

1. β- casein has a strong negatively charged N-terminal portion
2. At pH 6.6, the 21-residue N terminal sequence has a net charge of 12, while the remaining portion of the chain has almost no net charge

3. Extended helix formation is effectively prevented by the abundance of pro residues. Therefore, the β-casein molecule is a type of detergent that has a negatively charged head and a hydrophobic tail
4. The association between β-casein and temperature is notable for its strong dependence on both the presence and absence of Ca^{+2}
5. Only monomer is present at 44°C when Ca^{+2} is absent, but large polymers (20–24 monomers) form at room temperature.

γ-casein

1. It has long been known that the C terminal regions of the β casein sequence correspond to a group of caseins known as γ caseins. Plasmin, an enzyme, cleaves β casein at positions 28 to 29, 105, 106, and 107, 108 to form these
2. The γ caseins are made up of the fragments 29-209, 106-209, and 108-209. The smaller cleavage fragments are part of the whey's proteose-peptone fraction, which has been named for a long time. They are visible in the whey when acid precipitates casein. Thus fragments 1-105 and 1-107 were called as whey component 5, fragment 1-28 is whey component 8-fast, and fragments 29-105 and 29-107 were named as whey component 8-slow

κ-casein

1. About one-third of the κ casein molecules are carbohydrate free and contain only one phosphate group (SerP 149)
2. They differ in the quantity of N acetylneuraminic acid (NANA) residues they contain, and one of them, at least, seems to have an additional phosphate (SerP 127)
3. There are three distinct glycosyl oligomers connected to Thr 133 that have been found
4. Glutamic acid is the N terminal residue of κ casein
5. A mixture of polymers, most likely held together by intermolecular disulfide bonds, makeup κ casein

Casein Micelle Structure

Casein is present in the colloidal state. Numerous scientists, including Schmidt (1982), McMohan and Brown (1988), Holt (1992), Farrel (1983), Rollema (1992), and Visser (1993), have proposed various models of casein micelle structure.

Three broad categories can be applied to the suggested models:

- Core coat
- Internal structure
- Sub micelle model and the latest model is sub micelle/sub unit model

It was believed that submicelles of mass with a molecular weight of roughly 106 Dalton make up casein micelles. The submicelle is 10–15 nm in diameter. Mohr first proposed the submicelle model in 1967. He proposed that colloidal calcium phosphate (CCP) serves as a link between submicelles. The micelle's open porous structure is a result of this CCP. Hydrophobic and hydrophilic components primarily contribute to the stability of the micellar structure.

Several changes have been made to the submicelle model by Scmidt (1982), Walstra, and Jenness (1984). According to the submicelle model, the current structure of the casein micelle (Fig. 2) is that the submicelle's κ-casein content varies and that the κ-casein-deficient submicelle is found inside the micelle. The micelle has a κ-casein rich layer due to the concentration of the κ-casein rich submicelle at the surface, but some αs1, αs2, and β-casein are exposed on the surface as well. κ-casein's hydrophobic C terminal sticks out from the surface to form a layer that is 5–10 nm thick and gives the micelles a hairy look. By significantly influencing zeta potential and stearic stabilisation, these hairy layers are in charge of micellar stability. Micelles lose their colloidal stability when the hairy layer is eliminated, such as when ethanol collapses them or κ-casein is hydrolysed specifically, causing them to coagulate and precipitate.

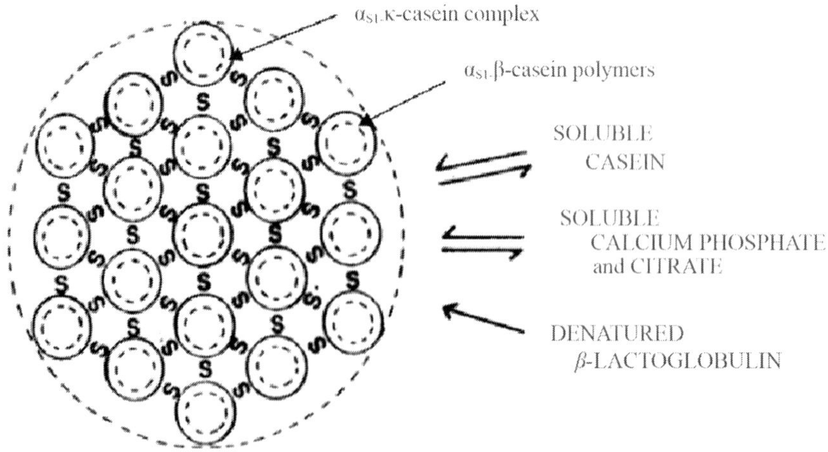

Fig. 2: Submicelle model of the casein micelle (adapted from Wong, 1988).

Despite explaining the majority of the physiochemical reactions that the casein micelle undergoes, the submicelle model has never received widespread acceptance. Recently, two more alternative models have been put forth. According to Visser (1992), the micelles are spherical conglomerates of individual casein molecules that are randomly aggregated and held together by a combination of forces, including a hydrophobic bond in the κ-casein surface layer and a salt bridge in the form of amorphous calcium phosphate.

The casein micelle was defined by Holt (1992, 1994) as a tangled web of flexible casein molecules forming a gel-like structure, from which the C terminal region of κ-casein extends to form a hairy layer, and where CCP microgranules are an integral feature. Two of the key components of the submicelle model are maintained in the Visser and Holt models: (1) CCP's cementing function and (2) κ-casein's primarily surface layer.

References

Arnold, R., K. M. Pruitt, M. F. Cole, J. M. Adamson, and J. R. McGhee. "Salivary antibacterial mechanisms in immunodeficiency,"in Saliva and DentalCarries, eds.I. Kleinberg, S. A. Ellison, I. D. Mandel(Washington: Information Retrieval Inc., 1979), 449–462.

Fox,P.E. and P.L.H. McSweeney, 1998. Dairy Chemistry and Biochemistry. Madras: Blackie Academic & Professional, Chapman & Hall.

Jenness,R. and S. Patton. Principles of Dairy Chemistry. New York: John Wiley & Sons, Inc., 1959.

Korhonen, H."Antimicrobial Factors in Bovine Colostrum."Journal of the Scientific Society of Finland 49 (1977):434–447.

Levay, P. F. and M. Vilijoen. "Lactoferrin: A General Review."Haematologica 80 (1995): 252–267.

Mathur, M.P., D. Datta Roy, and P. Dinakar. Text Book of Dairy Chemistry. New Delhi: Indian Council of Agricultural Research Govt. of India, 1999.

Webb, B.H., A. Johnson, and J. A. Alford. Fundamentals of Dairy Chemistry. West Port: The AVI Publ.Co.Inc., 1978.

Wong NP. 1988. "Fundamental of Dairy Chemistry." 3rd Edition. New York: Van Nostrand Reinhold: 481–492.

3

Milk Carbohydrate

Carbohydrates are polyhydroxy derivatives of aldehydes or ketones and undergo all the organic reactions common to the carbonyl groups and hydroxyl groups. The principal milk carbohydrate is lactose. It is a disaccharide. Hydrolysis of lactose with enzyme β-galactosidase yields glucose and galactose.

Lactose is the major carbohydrate present in all types of milk. The first record of isolation of lactose was in 1663, by Bartolettus by evaporation of whey. Milk contains only trace amounts of other sugars, including glucose, fructose, glucosamine, galactosamine, neuraminic acid, and neutral and acidic oligosaccharides.

Lactose plays a significant role in milk and milk products that are

- It plays a major role in the production of fermented dairy products.
- It contributes to the nutritive value of milk and milk products
- It affects the texture of certain concentrated and frozen products;
- It is involved in heat-induced changes in the color and flavor of highly heated milk products.

Status of Lactose

Lactose content varies extensively in the milk of different species. Breed of animal, individuality factors, udder infection, and especially the stage of lactation regulates the lactose content in milk. The lactose concentration in milk decreases gradually and significantly during lactation; this behavior contrasts with the lactational trends for lipids and proteins, which, after a decrease during early lactation, increase significantly during the second half of lactation. Lactose synthesis is hampered during mastitis. Lactose, along with sodium, potassium, and chloride ions, plays the principal role in controlling the osmotic pressure of the mammary system. Thus, any change in lactose content is compensated by a change in the soluble salt constituents. The osmotic relationship between lactose and soluble salts during the secretion of milk partly explains the presence of high lactose in milk with a reduced ash content and vice versa (Table 1).

Table 1: Lactose concentration of milk of some common species.

Species	Lactose content (%)
Goat	4.1
Camel	3.7
Cow	4.8
Sheep	4.8
Water buffalo	4.8
Human	7.0

Source: Miscellaneous.

Physico-Chemical Properties of Lactose

Structure of Lactose

Lactose is a disaccharide comprising galactose and glucose, linked by a β-l-4 glycosidic bond. Its systematic name is β-0-D-galactopyranosyl-(1-4)-α-D-glucopyranose (α-lactose) or β-0-D-galactopyranosyl-(1-4)-β-D-glucopyranose (β-lactose). The hemiacetal group of the glucose moiety is potentially free to make lactose a reducing sugar. Lactose exists in milk both as α- or β-anomer. The hydroxyl group on the C1 in the α form of glucose is cis to the hydroxyl group at C2 (oriented downward) (Fig. 1).

Proof of Lactose Structure

The structure and systematic name of lactose discloses four points.

1. The reducing group lies at no. 1 carbon atom of glucose.
2. It is a β-galactoside.
3. Both glucose and galactose have a pyranose ring structure.
4. Linkage is (1–4) linkage.
5. It is a disaccharide.

1. On oxidation lactose yields D-gluconic acid and galactose. During the reaction free aldehyde group of glucose moiety of lactose has been oxidized to a carboxylic acid group with the resultant production of D-gluconic acid. As the galactose moiety of lactose did not have a potentially free aldehyde group it escaped the oxidation. This suggests that the reducing group of lactose lies in the number one carbon atom of the D-glucose moiety of lactose.

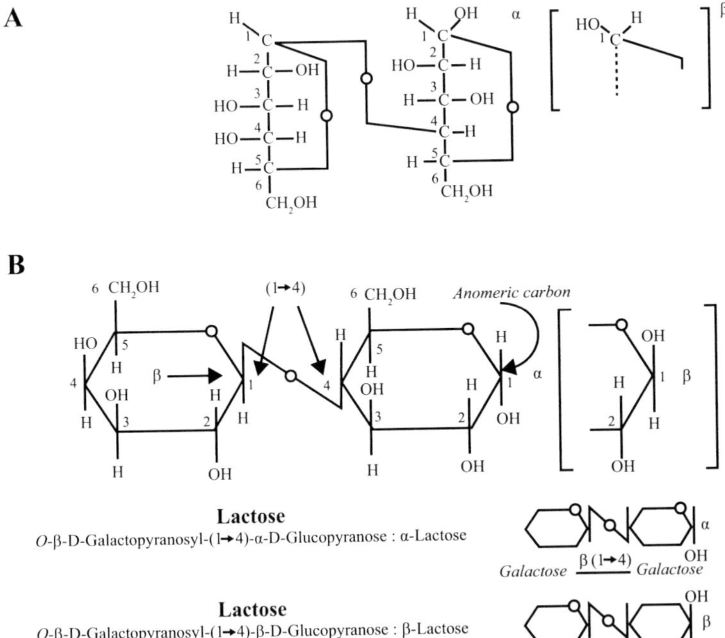

C

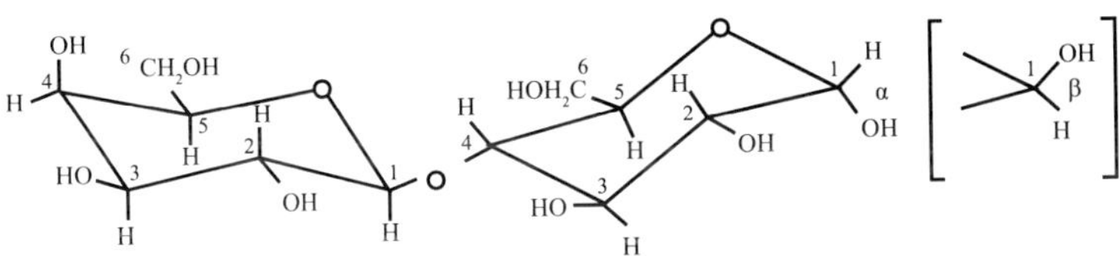

Fig. 1: α- and β-lactose. (A) Fischer projection, (B) Haworth projection, and (C) conformational formula.

Source: Fox and Mcsweeney (1998), Dairy Chemistry and Biochemistry, Blackie Academic and Professional, p. 24.

2. β-galactosidase hydrolyzes lactose to D-glucopyranose and D-galactopyranose. From the above considerations, we concluded that lactose is a β-galactoside.
3. On subsequent methylation of glucose and galactose moieties of lactose it is observed that no methylation has taken place at the fifth carbon atom of D-glucose and D-galactose. Therefore, it is concluded that both glucose and galactose have a pyranose ring structure.

4. On exhaustive methylation, it was found that there is no methylation at the number one carbon and number five carbon atom of D-galactose moiety, and in the case of glucose there is no methylation at the number one, fourth, and fifth carbon atom. Therefore, it manifests that the linkage between the D-glucose and D-galactose moiety of lactose is 1–4.
5. On hydrolysis lactose gives glucose and galactose. It proves that lactose is a disaccharide.

Physical Forms of Lactose

Lactose is found in milk either in α-lactose or β-lactose or amorphous glass form (mixture of both α and β forms). Lactose may produce several other forms when subjected to some special treatment.

α-Lactose Monohydrate ($C_{12}H_{22}O_{11}H_2O$)

Lactose crystallizes to α-monohydrate when crystallization is done below 93.5 °C. Precipitation of α-anhydrous form does not take place in milk. Crystallization of supersaturated solution at 93.5 °C or below gives α-monohydrate form. The same process gives β-anhydrous at a temperature above 93.5 °C. α-monohydrate contains 5% crystallization water and also has a solubility of 7 g/100 g water at 20 °C and specific rotation +89.4°. The melting point of α-monohydrate is −202 °C. Different crystal shapes of α-monohydrate are formed depending on the crystallization conditions. Tomahawk-shaped is the most common shape of α-monohydrate formed. Hard crystals are generally formed which dissolve slowly in water (Fig. 2).

Regular (Unstable) Anhydrous α-Lactose

Anhydrous α-lactose is prepared by the dehydration of α-hydrate in *vacuo* at a temperature ranging between 65 and 93.5 °C. This form is generally stable when no moisture is present.

$$\alpha\text{-hydrate} \xrightarrow[\text{vacuum}]{>100\,^{\circ}\text{C}} \alpha\text{-lactose anhydrous}$$

Moisture loss during the reaction is minimal up to 85 °C whereas significant moisture loss takes place at 90 °C which is more rapid at 120–125 °C.

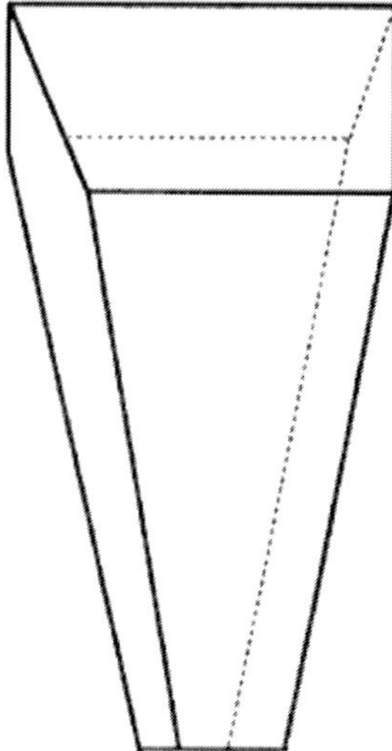

Fig. 2: Crystalline form of α-lactose monohydrate

Source: Fox and Mcsweeney (1998), Dairy Chemistry and Biochemistry, Blackie Academic and Professional, p. 30.

Anhydrous α-lactose is stable in dry air and is highly hygroscopic. In the presence of moisture, it forms α-hydrate form before dissolving. In a solution that is saturated with α-hydrate lactose then it will not dissolve this behavior suggests that no change has occurred in its properties.

Stable Anhydrous α-Lactose

The stable form of α-lactose is prepared by dehydrating α-hydrate in vacuum of 6–80 mm of Hg at temperatures between 100 to 190 °C and crystallizing it with suitable solvents especially dry methanol or 95% absolute ethanol.

$$\alpha\text{-hydrate} \xrightarrow[6-80\,\text{mmHg}]{100-190\ ^\circ\text{C}} \alpha\text{-lactose anhydrous}$$

The new crystalline structure reduces the tendency to absorb water so this anhydrous form remains stable even in an environment of 50% relative humidity. Products containing 99% anhydrous lactose can be prepared by refluxing α-hydrate crystals with methanol in the ratio of 1:10 for 1 h. At room temperature, the reaction is slower. The rate decreases with increasing proportion of α-hydrate in dry methanol. Anhydrous α-lactose is stable having high density, non-hygroscopic. It dissolves in water forming a hydrate. It dissolves readily in solution and is saturated with α-lactose hydrate. It is more soluble in water than either α-hydrate or β-anhydride form.

α and β Anhydrous Forms (Lactose Glass)

When a lactose solution is dried rapidly, viscosity increases so quickly that crystallization is impossible. A non-crystalline form is produced containing α- and β-forms in the ratio at which they exist in solution. Lactose in spray-dried milk exists as a concentrated sirup or amorphous glass which is stable if protected from air but is very hygroscopic and absorbs water rapidly from the atmosphere, becoming sticky. Upon shaking finely powder form of hydrated α-lactose with 10 times its weight with ethanol containing 1–5% anhydrous HCl, α-anhydrous lactose crystallizes out into tiny needle shape crystals in α and β anhydrous in a 5:3 ratio.

β-Anhydride Form

Since β-lactose is less soluble than the α-isomer above 93.5 °C, the crystals formed from aqueous solutions at temperatures above 93.5 °C are p-lactose; these are anhydrous and have a specific rotation of 35°. β-Lactose is sweeter than α-lactose but is not appreciably sweeter than the equilibrium mixture of a- and p-lactose normally found in solution (Table 2).

Table 2: Physical properties of the two common forms of lactose.

Property	**α-hydrate**	**B-anhydride**
Melting point (°C)	202	252
Specific rotation $[\alpha]_D^{20}$	+89.4	+35
Solubility in water ($g100\ mL^{-1}$) at 20 °C	7	50
Specific gravity (20 °C)	1.54	1.59
Specific heat	0.299	0.285
Heat of combustion ($kJ\ mol^{-1}$)	5687	5946

Source: Fox and Mcsweeney (1998), Dairy Chemistry and Biochemistry, Blackie Academic and Professional, p. 31.

Mutarotation

Sugars, having asymmetric carbon atoms, are optically active; their solutions rotate the plane of polarization of polarized light wave that passes through them either clockwise (dextrorotation, +) or anti-clockwise (laevorotation, –). Each sugar has its characteristic specific rotation, which is the angle, either positive or negative, by which the plane is rotated per unit concentration and unit path length.

In lactose, the configuration around the C_1, of glucose (i.e. the anomeric C) is not stable and can readily change (mutarotate) from the α- to the β-form and vice versa. This happens when the sugar is in solution because the hemiacetal form is in equilibrium with the open-chain aldehyde form which can be converted into either of the two isomeric forms (α or β). α- and β-lactose differ in specific rotation. A freshly made solution of either will change in rotation with time as equilibration with the other form occurs. This change in ratio and the conversion of one form to the other in solution is called mutarotation. These changes may be followed by measuring the change in optical rotation with time until, at equilibrium, the specific rotation is +55.4°.

	Specific rotation [α]D20
α-form	+89.4°
β-form	+35.0°
Equilibrium Mixture	+55.4°

Thus, the equilibrium mixture at 20 °C is composed of 62.7% β- and 37.3% α-lactose. The equilibrium constant, β/α, is 1.68 at 20 °C. The proportion of lactose in the α-form increases as the temperature is increased and the equilibrium constant consequently decreases. The equilibrium constant is not influenced by pH, but both temperature and pH affect the mutarotation. The change from α- to β-lactose is 51.1, 17.5 and 3.4% complete at 25, 15, and 0 °C, respectively, in 1 h and is almost instantaneous at about 75 °C. The rate of mutarotation is slowest at pH 5.0, increasing rapidly at more acid or alkaline values; equilibrium is established in a few minutes at pH 9.0. The presence of sugar and salt also affects mutarotation. Salt increases the mutarotation while sugar decreases the mutarotation. The specific rotation of lactose varies with the solvent. Lactose has specific rotation with Glycerol, water, alcoholic, or acidic solution in decreasing order.

Solubility

The solubility characteristics of the α- and β-isomers exhibit wide variations. When α-lactose is added in excess to water at 20 °C, about 7 g per 100 g of water dissolves immediately. Some α-lactose mutarotates to the β anomer to establish the equilibrium ratio 62.78β:37.3α; therefore, the solution becomes unsaturated concerning α and more α-lactose dissolves. These two processes (mutarotation and solubilization of α-lactose) continue until two criteria are met i.e. ~7 g α-lactose in solution and a β/α ratio of 1.6:1.0. Since the β/α ratio at equilibrium is about 1.6 at 20 °C, the final solubility is 7 g + (1.6 × 7) g = 18.2 g per 100 g water.

When β-lactose is dissolved in water, the initial solubility is ~50 g per 100 g water at 20 °C. Some β-lactose mutarotates to α to establish a ratio of 1.6:1. At equilibrium, the solution would contain 30.8 g β and 19.2 g α/100 mL; therefore, the solution is supersaturated with α-lactose, some of which crystallizes, upsetting the equilibrium and leading to further mutarotation of

$\beta \rightarrow \alpha$. These two events, i.e. crystallization of α-lactose and mutarotation of β, continue until the same two criteria are met, i.e. ~7 g α-lactose in solution and a β/α ratio of 1.6:1. Again, the final solubility is 18.2 g lactose per 100 g water. Since β-lactose is much more soluble than α and mutarotation is slow, it is possible to form more highly concentrated solutions by dissolving β- rather than α-lactose. In either case, the final solubility remains the same.

The solubility of lactose as a function of temperature is summarized below. The solubility of α-lactose is more temperature dependent than that of β-lactose and the solubility curves intersect at 93.5 °C. A solution at 60 °C contains approximately 59 g lactose per 100 g water. Suppose that a 50% solution of lactose (30 g β- and 20 g α-) at 60 °C is cooled to 15 °C. At this temperature, the solution can contain only 7 g α-lactose or a total of 18.2 g per 100 g water at equilibrium. Therefore, lactose will crystallize very slowly out of solution as irregularly sized crystals which may give rise to a sandy, gritty texture (Fig. 3).

Crystallization of Lactose

The solubility of lactose is temperature-dependent and solutions are capable of being highly supersaturated before spontaneous crystallization occurs, and even then, crystallization may be slow. In general, super solubility at any temperature equals the saturation (solubility) value at a temperature of 30 °C or higher. The insolubility of lactose, coupled with its capacity to form supersaturated solutions, is of considerable practical importance in the manufacture of concentrated milk products.

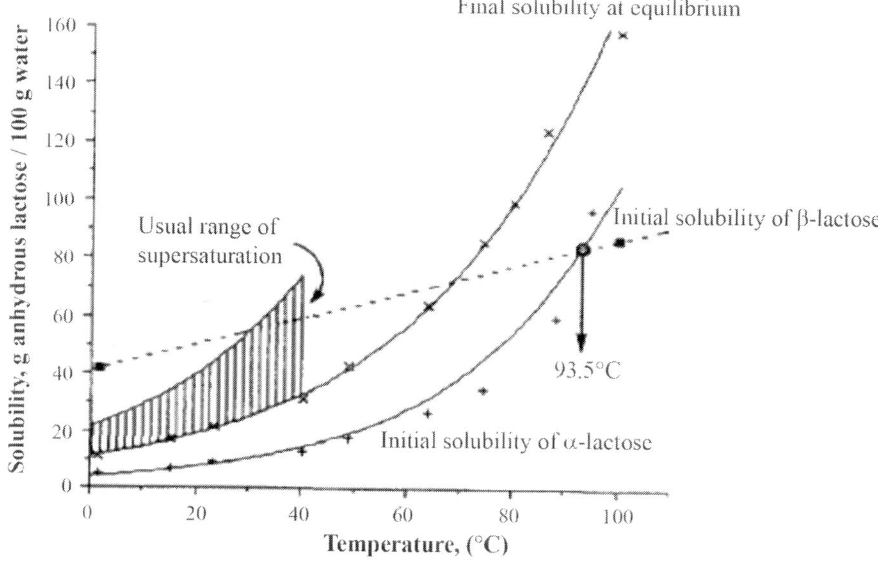

Fig. 3: Solubility of lactose in water.

Source: Fox and Mcsweeney (1998), Dairy Chemistry and Biochemistry, Blackie Academic and Professional, p. 28.

Solutions of lactose are capable of being highly supersaturated when nuclei and agitation are absent before the spontaneous crystallization. Even in such solutions, crystallization occurs. Solubility curves for lactose are shown in the figure and are divided into unsaturated, metastable, and labile zones. Cooling a saturated solution or continuing concentration beyond the saturation point leads to supersaturation and produces a metastable area where crystallization does not occur readily. At higher levels of supersaturation, a labile area is observed where crystallization occurs readily. The pertinent points regarding supersaturation and crystallization are:

- The unsaturated region does not show any nucleation or crystal growth.
- Crystal growth can occur in the metastable as well as in the labile areas.
- Nucleation occurs in metastable areas only if seeds (center for crystal growth) are added.
- Spontaneous crystallization can occur in the labile area without the addition of seeding material (Fig. 4).

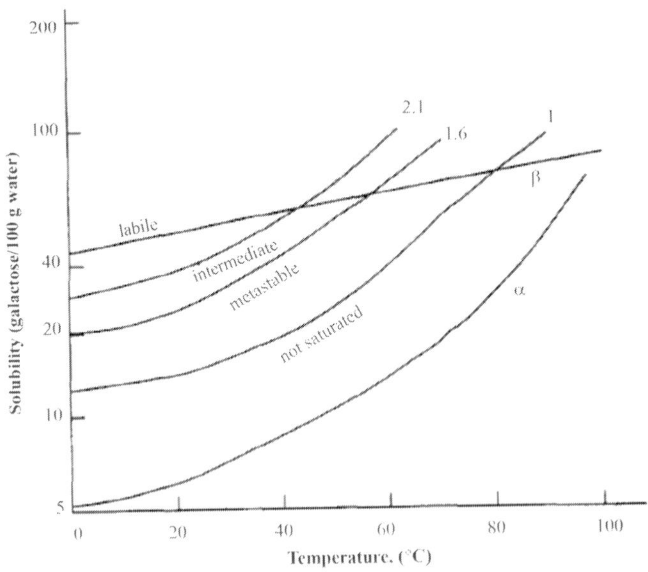

Fig. 4: Initial solubility of α-lactose and β-lactose, final solubility at equilibrium (line l), and supersaturation by a factor 1.6 and 2.1 (αlactose excluding water of crystallization).

Source: Fox and Mcsweeney (1998), Dairy Chemistry and Biochemistry, Blackie Academic and Professional, p. 29.

The rate of nucleation is slow at low levels of supersaturation and in highly supersaturated solutions due to the high viscous nature of the solution. The stability of a lactose "glass" developed because of the low probability of nuclei forming at very high concentrations. Once a sufficient number of nuclei have formed, crystal growth occurs at a rate influenced by:

- degree of supersaturation;
- surface area available for deposition;
- viscosity;
- agitation;
- temperature;
- mutarotation, which is slow at low temperatures.

Crystal shape depends upon the conditions of crystallization. Principle factor which governs the crystal shape is precipitation pressure. If the precipitation pressure is high i.e. concentration of lactose is high the crystallization is forced to high speed and only prism crystals are formed. On reduced precipitation pressure diamond or plate-shaped crystals are formed. On down, further down, and below precipitation pressure crystals of the pyramid, tomahawk, and fully developed crystals of free size and angle are formed respectively. Recent studies on the growth rate of lactose crystals show that the rate of crystallization is high if the precipitation pressure is high. Lactose crystal once formed grows only in one direction of the principle axis.

Impurities in milk inhibit lactose crystallization by retarding nucleation. It gives crystals of irregular, clump shape.

Effect of Additives on the Growth Rate of Crystallization

Few additives are found to inhibit the growth while some accelerate the crystallization. The effect of additives was studied on all the faces and it was retarding growth on certain faces whereas promoting growth on certain faces. Gelatin can retard crystallization by 1/3 to ¾ to that of normal even at low concentrations. But in highly supersaturated solution gelatin has failed to inhibit sandiness in ice cream. Present-day marine and vegetable gums are used in ice cream formulations because they have reduced instances of lactose crystallization.

Effect of Carbohydrates

Different carbohydrates actively inhibit lactose crystallization, and other don't. Carbohydrate containing β-galactosyl group or 4-0-glucose inhibit the lactose crystal growth. β-lactose which is always formed in any solution containing lactose has been presented to retard the crystallization process greatly. The retarding action is ascribed to the fact that the β-galactoside part of the molecule is that of lactose. The β-galactoside molecule get attached to α-molecule which are acceptors of the β-galactosyl group. Once incorporated in crystal they retard further growth because of their β-glucose group.

Effect of Riboflavin

It may also be adsorbed on growing lactose crystals and alter the crystallization hand. Adsorption depends upon (1) concentration of riboflavin, (2) concentration of lactose, (3) temperature. There is no adsorption below 2.5 μg/mL concentration of riboflavin whereas adsorption increases linearly as concentration increases. If the concentration of lactose is higher the crystallization is faster and riboflavin will not be able to adhere on site. An increase in temperature makes it in more soluble form so it will not be available for adsorption. By proper control, 200–300 μg rib./g lactose is the optimum concentration for lactose crystallization.

Effect of Methanol or Ethanol

It increases crystallization at low concentrations. Alcohol might promote nucleation.

Effect of pH

The rate of crystallization increases at low pH around 1. The effect of low pH might influence crystal surface reaction.

Sweetness

Lactose is less sweet as compared to other common sugars as sucrose, fructose, and glucose. It is relatively sweeter at higher concentrations. Thus, for example, 1, 5, 10, and 20% sucrose solutions are equivalent in sweetness to 3.5, 15, 20, and 33% lactose solutions. β-lactose is sweeter than α-lactose, but this difference is not important since the small difference between freshly prepared solutions is eliminated quickly by equilibration of the anomers.

Chemical Properties of Lactose

Carbohydrates undergo extensive, complex chemical changes under different conditions. Lactose undergoes different reactions which involve the following four sites of attack:

a) the 1–4-linkage between the galactose and glucose units,
b) the reducing group of the glucose unit,
c) the hydroxyl groups of both the glucose and galactose units; and
d) the carbon-to-carbon bonds.

Hydrolysis

Hydrolysis of lactose to glucose and galactose may be affected by the following agents: mineral acids, lactase enzyme preparations, and ion exchange resins. Organic acids, such as citric, which satisfactorily hydrolyze sucrose are largely without effect on lactose. Stronger mineral acids, such as hydrochloric and sulfuric acid are necessary for the hydrolysis of lactose. In general, higher temperatures and higher concentrations of acid favor more rapid hydrolysis. However, such conditions also favor undesirable side reactions such as discoloration, bitter and odors substances, formation of furan compounds, and other sugar fragments. A 10% lactose solution adjusted to pH 1.2 with hydrochloric acid is hydrolyzed to glucose and galactose when held at 150°C for 1 h. At laboratory conditions, the same reaction can be carried by hydrolyzing 5% solution at 90°C by treating it with 10 mL of concentrated hydrochloric acid per 100 mL of solution for 90 min.

$$\text{Lactose} \xrightarrow{H^+} \text{Glu cos e} + \text{Galactose}$$

$$\begin{matrix}\text{10\% Lactose} \\ \text{(Aqua solution)}\end{matrix} \xrightarrow[150]{\text{pH1.2 with HCl}} \text{Glu cos e} + \text{Galactose}$$

$$\begin{matrix}\text{5\% Lactose} \\ \text{(Aqua solution)}\end{matrix} \xrightarrow[\text{10ml HCl/100sol.}]{\text{90oC/90min}} \text{Glu cos e} + \text{Galactose}$$

Hydrolysis of lactose may be done mainly by three significant origins of lactase enzymes. They are: (a) certain species of yeasts, e.g. *Torula cremoris*; (b) the intestinal mucosa of mammals, particularly that of the calf; and (c) β-galactosidase from almonds. Of these, the lactase preparations from yeast appear to be the most important. Commercially available lactase from yeast shows optimum activity between pH 6.0 and 6.5 and an incubation temperature of 45°C. This lactase is inactivated by heating to 75°C for 15 min. Enzyme concentrations between 1 and 5%, based on lactose will hydrolyze about seven times their weight of lactose per hour at 40 °C. Hydrolysis proceeds rapidly to 50% of total and more slowly to 70%. Activity beyond 75% hydrolysis is very low. Yeast lactose has been observed to act satisfactorily in whole milk, skim milk, whole and skim milk concentrates, sweetened condensed milk, and whey.

The hydrolysis of lactose in aqueous solution has been achieved satisfactorily with sulfonated polysterene resins in the hydrogen form. Since sulfonic acids are strong acids, hydrolysis by them does not differ in principle from the action of mineral acid. However, discoloration of lactose solution and certain other undesirable side reactions can be avoided by ion exchange resin hydrolysis.

Pyrolysis

Pyrolysis is the chemical decomposition of lactose by the action of the heat. Vapor pressure measurements indicate that water may be lost at 85°C and even at 80°C (10.4 mm of Hg) under very low humidity. After several days at 100°C the water of crystallization is almost entirely lost from the powder. At 130°C the water is lost quickly, giving rise to an anhydrous powder. At 150° to 165°C lactose becomes yellow and at 175°C it becomes brown, emits a characteristic odor, and loses about 13% of its original weight. The most important heat-induced changes in dairy products that involve lactose are the changes associated with browning.

Oxidation

Oxidation of lactose will vary with the particular reagent, its concentration, and other conditions of the reaction. Thus, by selection of conditions, it is possible to derive oxidation products from lactose, which range from relatively simple alteration of the reducing carbon in the glucose portion of the molecule to a carboxylic acid group, to complete degradation, with the end products being CO_2 and water. The conversion of lactose to lactobionic acid or, more properly, 4-(β-D-glactopyraosyl)-D-gluconic acid is considered the most significant among the reactions of lactose. Reagents that favor the complete oxidation of lactose to CO_2 and water are acid or alkaline potassium permanganate; alkaline conditions with catalytic amounts of ferrous hydroxide, ferrous sulfate, and sodium sulfite; and sunlight with zinc oxide serving as catalyst.

Lactose can be oxidized to lactobionic acid by mild chemical dehydrogenation and by lactose dehydrogenase produced by certain species of bacteria of the genus *Pseudomonas*. Vigorous oxidation of lactose with dilute nitric acid ruptures the glycosidic linkage and produces dicarboxylic acid derivatives of the two sugars. The dicarboxyl derivative of galactose, galactaric or, mucic acid, was formerly much used as a crystalline derivative for identification of galactose.

Biological oxidation of lactose to CO_2 and water is brought about by mixed cultures of bacteria and protozoa obtained from sewage sludge. Such processes are useful in decomposing lactose-containing wastes from dairy factories.

Reduction

Only the free aldehyde of lactose reacts in reduction to become an alcohol group, and lactitol (4-0-β-D-galcatopyranosyl-D-glucitol) is formed. Using certain conditions of pressure and temperature it is possible to hydrogenate (reduce) lactose to lactositol. An 80% yield has been reported under such conditions, using a temperature of 145 °C and a pressure of approximately 120 atmospheres. When conditions favor the hydrolysis of the lactose or lactitol, hexitols (dulcitol and sorbitol) are obtained. Under highly rigorous conditions of reduction, various other monohydric and polyhydric alcohols are obtained.

References

Fox, P. E. and P. L. H. McSweeney. Dairy Chemistry and Biochemistry. Madras: Blackie Academic & Professional and Chapman & Hall, 1998.

Jenness, R. and S. Patton. Principles of Dairy Chemistry. New York: John Wiley & Sons, Inc., 1959.

Mathur, M. P., D. Datta Roy, and P. Dinakar. Text Book of Dairy Chemistry. New Delhi: Indian Council of Agricultural Research Govt. of India, 1999.

Webb, B. H., A. Johnson, and J. A. Alford. Fundamentals of Dairy Chemistry. West Port: The AVI Publ. Co. Inc., 1978.

4

Milk Lipid

Milk lipids or milk fats exist as an emulsion of small spherical droplets called milk fat globules. These globules are very small in size but very large in number. The diameter of the milk fat globules ranges between 0.1 and 22 μm and the number varies between 1.5 and 3.0 billion globules/mL. The average size of milk fat globules ranges from 2 to 5 μm. The average fat globule size in cow and buffalo milk are 3.4 and 4.5 μm, respectively. The surface of the milk fat globules is coated with an adsorbed layer of material known as the milk fat globule membrane (MFGM). MFGM contains phospholipids and proteins in a complex form. MFGM has two main functions (a) fat globule stabilization and (b) preservation of the fat globule's identity.

General Composition

Three distinctly different phases of milk lipid or fat, namely, the fat globules, the membrane surrounding these globules, and the serum, are generally observed (Table 1).

Table 1: General composition of milk lipid.

Constituent	Percentage
Triglycerides	95–96
Diglycerides	1.26–1.59
Keto acid glycerides	0.85–1.28
Ketogenic glycerides	0.03–0.13
Hydroxy acid glycerides	0.6–0.78
Lactogenic glycerides	0.06
Neutral glyceryl ethers	0.016–0.020
Neutral plasmalogens	0.04
Free fatty acids	0.1–0.44
Phospholipids (total)	0.80–1.00
Sphingolipids	0.06
Sterols	0.22–0.41
Squalene	0.007
Carotenoids	0.007–0.009

Constituent	Percentage
Vitamin A	0.0006–0.0009
Vitamin D	0.00000085–0.00000021
Vitamin E	0.0024
Vitamin K	0.0001

Source: Miscellaneous.

Triacylglycerols (triglycerides) represent 97–98% of the total lipids in the milk of most species. The diglycerides probably represent incompletely synthesized lipids in most cases. Although phospholipids represent less than 1% of total lipids, they play an important role in being present mainly in the MFGM and other membranous material in milk. The principal phospholipids are phosphatidylcholine, phosphatidylethanolamine, and sphingomyelin. Trace amounts of other polar lipids, including ceramides, cerebrosides, and gangliosides, are also present. Phospholipids represent a considerable percentage of the total lipid of buttermilk and skim milk reflecting the presence of proportionately larger amounts of membrane material in these products.

Cholesterol is the principal sterol in milk (>95% of total sterols); the level (~0.3%, w/w, of total lipids) is low compared with many other foods. Most of the cholesterol is in free form, with less than 10% as cholesteryl esters. Several other sterols, including steroid hormones, occur at trace levels. Several hydrocarbons occur in milk in trace amounts. Of these, carotenoids are the most significant. In quantitative terms, carotenes occur at only trace levels in milk (typically ~200 $\mu g/L^{-1}$) but they contribute 10–50% of vitamin A activity in milk. Carotenes are responsible for the yellowish color of milk fat. The milk carotenoid content varies with breed and very markedly with season. Milk contains significant concentrations of fat-soluble vitamins, milk and dairy products make a significant contribution to the dietary requirements for these vitamins in Western countries. The actual form of the fat-soluble vitamins in milk appears to be uncertain and their concentration varies widely with breed of animal, feed, and stage of lactation. The vitamin A activity of colostrum is almost 30 times higher than that of mature milk. Several prostaglandins occur in milk but it is not known whether they play a physiological role; they may not survive storage and processing in a biologically active form. Human milk contains prostaglandins E and F at concentrations 100-fold higher than human plasma and these may have a physiological function, e.g. gut motility.

Milk fat provides richness and smoothness to dairy products. Freshly secreted milk contains microscopic emulsion of globular liquid fat which is present in the aqueous phase of milk plasma. Fat is considered to be the most variable milk component. Buffalo milk fat globules are larger (4.15–4.60 µm) than the fat globules present in cow milk (3.36–4.15 µm). Three classes of milk fat globules classified based on size are given in Table 2.

Table 2: Size distribution of milk fat globules.

Class	Diameter (µm)	The proportion of the total globule population (%)	Fraction of total milk lipid (%)
Small	Below 2	70–90	<5
Intermediate	3–5	10–30	90
Large	8–10	0.01	1–4

Source: http://ecoursesonline.iasri.res.in/

The milk fat globules are stabilized by a very thin membrane commonly named as Milk Fat Globule Membrane (MFGM). The thickness of the membrane varies between 5 and 10 µm. Proteins, lipoproteins, phospholipids, lipids, cerebrosides, enzymes, nucleic acids, trace elements, and bound water are the major components of the membrane (Table 3). The major function of the membrane is to keep the fat separated from the other components when it is exposed to external forces during the processing and handling of milk. It also shields the milk fat from the action of enzymes, mainly lipase which causes rancidity in milk and milk products. Alkaline phosphatase and xanthine oxidase are attached to the milk fat globule membrane along with minerals like iron and copper.

Table 3: Average compositions of the milk fat globules membrane.

Component	Buffalo	Western cow
Lipid content	38.5	35.7
Neutral lipids	74.4	65.7
Phospholipids	19.7	18.6
Triglycerides	52.7	45.5
Diglycerides	7.4	5.5
Sphingomyelin	4.3	3.2
Phosphatidyl serine	3.1	2.4
$C_{18:0 \text{ Stearic acid}}$	34.4	24.1
$C_{16:0 \text{ Palmitic acid}}$	22.4	23.8
$C_{18:1 \text{ Oleic acid}}$	14.5	14.5
$C_{14:0 \text{ Myristic acid}}$	12.6	14.4
Protein (g/100 g fat globule)	1.36	1.8

Source: http://ecoursesonline.iasri.res.in/

Fatty Acid Profile of Milk Fat

Triglycerides constitute 95–99% of milk fat. Other parts of milk fat are composed of diglycerides (approximately 4.1% in buffalo milk and 1.26–1.59% in cow milk), and monoglycerides (approximately 0.7% in buffalo milk and 0.016–0.038% in cow milk). High, medium, and low molecular weight triglycerides content of buffalo milk are 42.5, 17.1, and 40.5%, respectively. Cow milk fat contains 52.9, 18.9, and 28.2% of high, medium, and low molecular weight triglyceride respectively. Buffalo milk contains a lower amount of free fatty acid (0.22%) as compared to cow milk (0.33%). Fatty acids control the functional properties of milk fat. Approximately 65% saturated, 30% monounsaturated, and 5% polyunsaturated fatty acids are present in the milk fat. Nutritionally they function differently. Saturated fatty acids are closely associated with high blood cholesterol and are responsible for heart disease. However, the metabolism pattern of short-chain fatty acids (C4–C8) is different than long-chain fatty acids (C16–C18) and short-chain fatty acids are not held responsible for heart disease. Conjugated linoleic acid (CLA) present in milk fat is considered beneficial nutritionally. The short-chain fatty acids (C4–C8), medium-chain fatty acids (C10–C14) and long-chain fatty acids (C16 or higher) content of milk fat are 7, 15–20, and 73–78%, respectively. Detailed fatty acid profiles of cow and buffalo milk are given in Table 4.

Table 4: Fatty acid profile of buffalo and cow milk fat.

Fatty acids	Common name	Composition (wt/wt%)	
		Buffalo milk	Cow milk
$C_{4:0}$	Butyric	4.36	3.20
$C_{6:0}$	Caproic	1.51	2.11
$C_{8:0}$	Caprylic	0.78	1.16
$C_{10:0}$	Capric	1.28	2.57
$C_{10:1}$	Caproleic	-	0.31
$C_{12:0}$	Lauric	1.78	-
$C_{14:0}$	Myristic	10.81	11.93
$C_{14:1}$	Myristoleic	1.27	2.12
$C_{15:0}$	Pentadecanoic	1.29	1.23
$C_{16:0\ (branched)}$	-	0.18	0.30
$C_{16:0}$	Palmitic	33.08	29.95
$C_{16:1}$	Palmitoleic	1.99	2.16
$C_{17:0}$	Margaric	0.58	0.34

$C_{18:0\ (branched)}$	-	0.24	0.35
$C_{18:0}$	Stearic	11.97	10.07
$C_{18:1}$	Oleic	27.15	27.42
$C_{18:2}$	Linoleic	1.51	1.49
$C_{18:3}$	Linolenic	0.47	0.59

Source: http://ecoursesonline.iasri.res.in/

Buffalo milk fat contain higher amount of butyric acid than cow milk fat. However, Buffalo milk has lower amount of other short-chain fatty acids (caproic, caprylic, capric, and myristic). The long-chain fatty acids (palmitic and stearic) are comparatively higher in buffalo milk. The unsaturated fatty acid level is almost similar both in cow and buffalo milk. Different quantities of other lipids such as phospholipids, sterols, vitamins A, D, E, and K, carotenoids, and traces of free fatty acids are also present in milk fat.

Physico-chemical differences between the fats of buffalo and cow milk are summarized in Table 5.

Cholesterol

Cholesterol is a nutritionally important component of milk fat. Species, breed, feed, stage of lactation, and season of the year largely affect the cholesterol content. Western breeds of cattle contain the highest amount of cholesterol (317–413 mg/100 g fat), followed by zebu (desi) cow (303–385 mg/100 g fat). Cholesterol content is lower in buffalo milk (235–248 mg/100 g fat). Cholesterol level increases with the increase in the lactation period. Colostrum contains a very high amount of cholesterol (570–1950 mg per 100 g fat) in the first milking after parturition which is then gradually declined to the normal level after successive milking.

Phospholipids

The average phospholipid content of buffalo milk is 21.04 mg/100 mL of milk. The phospholipid content of cow milk is 33.71 mg/100 mL. Concentration of phospholipids fractions present in buffalo and cow milk is given in Table 6.

Table 6: The concentration of phospholipid fractions in buffalo and cow milk.

Phospholipid fraction	**Concentration (mg/100 mL)**	
	Buffalo milk	**Cow milk**
Lecithin	7.29	10.01
Cephalin	9.12	15.57
Sphingomyelin	4.63	8.13

Source: http://ecoursesonline.iasri.res.in/

Crystalline Morphology of Edible Fat

Edible fats generally originate from vegetable, marine, and animal sources. They all form three main types of crystal namely, a, b′, and b forms. The a form is the least stable and existence is usually transitory. It consists of fragile, translucent crystals 5 mm in length. The b′ crystals consist of tiny, delicate crystals of 1 mm in length and the b-form is relatively large and coarse crystals, which average 25–50 mm in length, but which may grow as large as 100 mm or more. An intermediate form has also been observed, which comprises crystals 3–5 mm in length which tend to aggregate in coarse clumps.

Liquid fats and triacylglycerols cooled without agitation tend to form a-crystals initially. Further cooling of the a form leads to tighter chain packaging and a gradual transition to the b-form cooling with agitation, which, however, favors the formation of the b′ form.

Although fat may exist in any of the three polymorphic forms, different fats tend to transform into either the b or b′ form. The polymorphic behavior is largely influenced by the composition of the fatty acids and the positional distribution of the acyl glycerols. Fats that consist of relatively few triacyl glycerols of similar structure tend to transform to b form, while heterogeneous fats transform more slowly. In most cases, fats that tend to transform into the b type contain low (10%) quantities of palmitic acid, while those that transform into the b′ type contain 20%.

Milk fat is of the b′ type. Vegetable fats commonly used in margarine and spreads, coconut, olive, peanut, soybean, and sunflower oil are of the b type, while rapeseed, cotton seed, and palm oil are b′ type.

β′ crystals are desirable for butter, margarine, and spreads and in the latter two cases, a serious defect called "graininess" may be overcome by using b-type fats as the "soft" component of the fat blend, although other conditions may be manipulated to favor the formation of b′ crystals. For this reason, co-randomization is used to enable margarine to be made from unblended sunflower oil without problems associated with b-crystal formation.

Lipid Oxidation

Lipid oxidation, leading to oxidative rancidity, is a major cause of deterioration in milk and dairy products. Lipid oxidation is an autocatalyzed free radical chain reaction that is normally divided into three phases

a. Chain initiation

b. Chain propagation

c. Chain termination

The initial step involves abstracting a hydrogen atom from a fatty acid, forming a fatty acid (FA) free radical, e.g.

$$CH_3\text{----}CH_2\text{–}CH\text{=}CH\text{–}CH\text{–}CH\text{=}CH\text{–}CH_2\text{------}COOH$$

Although saturated fatty acids may lose a H and undergo oxidation, the reaction principally involves unsaturated fatty acids, especially polyunsaturated fatty acids (PUFA), the methylene–CH_2–, group between double bonds being particularly sensitive:

$$C_{18:3} >> C_{18:2} >> C_{!8:1} > C_{18:0}$$

The polar lipid in milk fat is richer in PUFA than neutral lipids, is concentrated in the fat globule membrane with several pro-oxidants, and is therefore, particularly sensitive to oxidation.

The initiation reaction is catalyzed by singlet oxygen, polyvalent metal ions that can undergo a monovalent oxidation/reduction reaction, especially copper and light, in the presence of riboflavin, a photosensitizer.

The FA radical may abstract a H from a hydrogen donor, e.g. an antioxidant (AH), terminating the reaction, or it may react with molecular triplet oxygen, $^3O_2$, forming an unstable peroxy radical. In turn the peroxy radical may obtain an H from an antioxidant, terminating the reaction, or from another fatty acid, forming a hydroperoxide and other FA-free radical, which continues the reaction.

The intermediate products of lipid oxidation are themselves free radicals, and more than one may be formed during each cycle; hence, the reaction is autocatalytic. Thus, the formation of only a very few free radicals by an exogenous agent is necessary to initiate the reaction. The reaction shows an induction period, the length of which depends on the prooxidants and antioxidants.

The hydroperoxides are unstable and may break down into various products, including unsaturated carbonyls, which are mainly responsible for the off-flavors of oxidized lipids. The FA free radicals, peroxy radicals, and hydroperoxides are flavorless. Different carbonyls vary with respect to flavor impact, and since the carbonyls produced are dependent on the fatty acid being oxidized, the flavor characteristics of oxidized dairy products vary.

Factors of Lipid Oxidation

A. Potential pro-oxidants

1. Oxygen and activated oxygen species.
2. Riboflavin and light.

3. Metals (e.g. Copper and iron) associated with various ligands.

   Metallo proteins

   Salts of fatty acids

4. Metallo enzymes

   Xanthine oxidase, lactoperoxidase, catalase, cytochrome P420, sulfhydryl oxidase.

B. Potential antioxidants

1. Tocopherols.
2. Milk proteins.
3. Carotenoids.
4. Certain ligands for metal prooxidants.
5. Ascorbate and thiols.
6. Maillard browning reaction products.
7. Enzymes (superoxide di mutase, sulfhydryl oxidase)

C. Environment and physical factors

1. Inert gas or vacuum packing.
2. Gas permeability and opacity of packaging material.
3. Light.
4. Temperature.
5. pH.
6. Water activity.
7. Reduction potential.
8. Surface area.

D. Processing and storage

1. Homogenization.
2. Thermal treatments.
3. Fermentation.
4. Proteolysis.

Lipid Oxidation **(adapted from Fox and Mc Sweeney, 1998)**

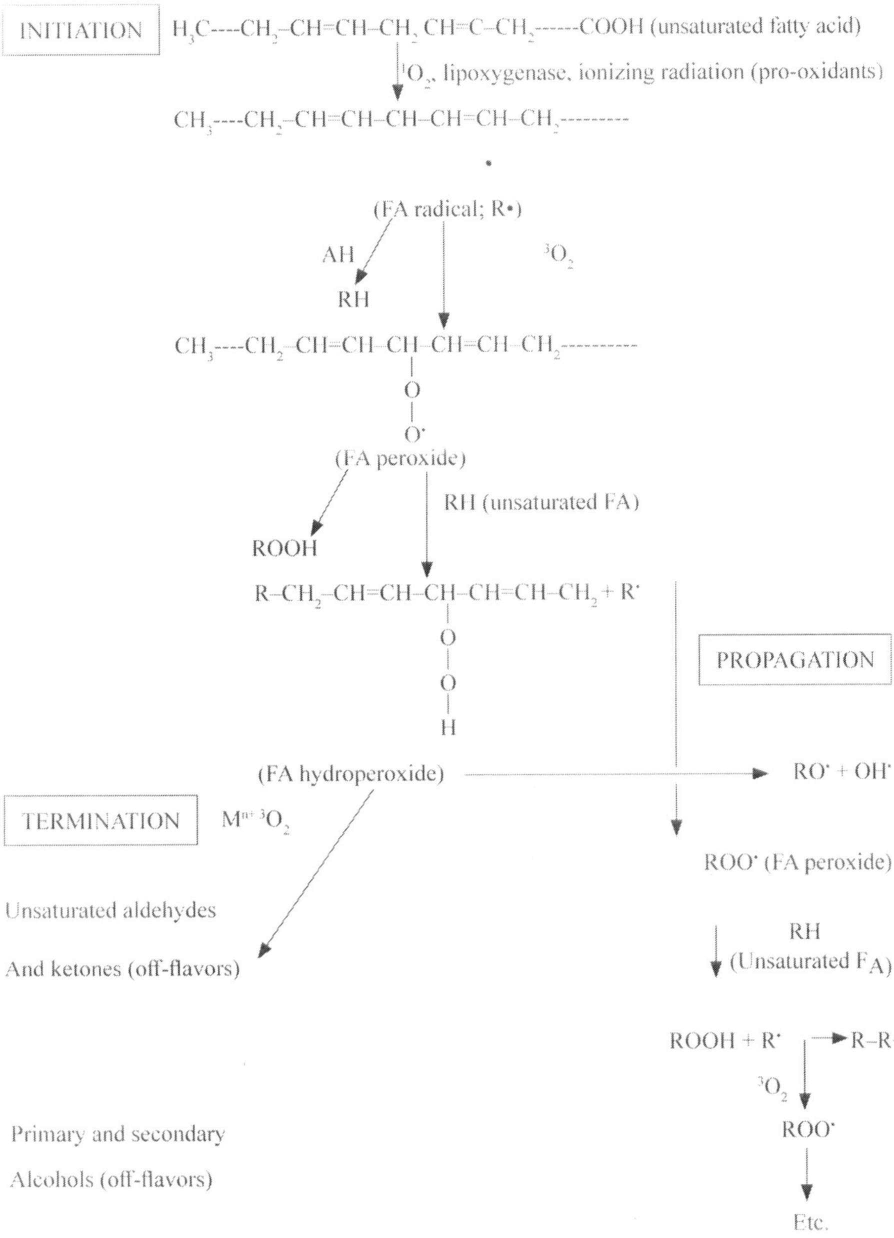

INITIATION H_3C----CH_2–CH=CH–CH_2 CH=C–CH_2------COOH (unsaturated fatty acid)

$^1O_2$, lipoxygenase, ionizing radiation (pro-oxidants)

CH_3----CH_2–CH=CH–CH–CH=CH–CH_2--------

(FA radical; R•)

AH → RH

$^3O_2$

CH_3----CH_2–CH=CH–CH–CH=CH–CH_2---------
with –O–O˙ on the central CH

(FA peroxide)

ROOH

RH (unsaturated FA)

R–CH_2–CH=CH–CH–CH=CH–CH_2 + R˙
with –O–O–H on the central CH

PROPAGATION

(FA hydroperoxide) → RO˙ + OH˙

TERMINATION M^{n+} $^3O_2$

ROO˙ (FA peroxide)

Unsaturated aldehydes

And ketones (off-flavors)

RH (Unsaturated FA)

ROOH + R˙ → R–R

$^3O_2$

ROO˙

Primary and secondary

Alcohols (off-flavors)

Etc.

Rancidity

Oil and fats begin to decompose when they are isolated from their natural living environment. The presence of free fatty acids is an indication of lipase activity or other hydrolytic changes. Changes occur during storage, which result in the production of an unpleasant taste and odor. Such oils and fats are referred to as having become rancid. The unpleasant organoleptic characteristics are in part caused by the presence of free fatty acids, but the major development of rancidity is brought about by atmospheric oxidation.

Type of rancidity

1. Hydrolytic rancidity.
2. Ketonic rancidity.
3. Oxidative rancidity or auto oxidation.

1. Hydrolytic rancidity

Hydrolytic rancidity resulting from the hydrolysis of fat to short chain volatile free fatty acids such as butyric acid. Lipase enzyme is responsible for this deterioration. The enzyme derived from the milk itself or they may be of microbial origin.Here, fatty acids are broken away from triglyceride, leaving behind di and mono glyceride or glycerol depending on whether 1, 2, or 3 fatty acid radicals have been hydrolyzed from each molecule.

Tri glyceride ⟶ 1, 2 di glyceride ⟶ 2 Mono glyceride ⟶ Glycerol
2, 3 di glyceride + Fatty acid | + Fatty acid | + Fatty acid

Milk lipase hydrolyze preferentially the primary ester position. Adequate high temperature is sufficient to destroy the lipase and thus prevents hydrolytic rancidity.

2. Ketonic rancidity

Ketonic oxidation produce carbonyl compound especially, methyl ketone. Here lower fatty acids such as butyric acid is utilized and converted into ketone with strong odor. The transformation of butyric acid to acetone by *Aspergillus niger* took place via a hydroxy acid.

$$R–CH_2–CH_2–COOH$$
$$\downarrow -2H$$
$$R–CH=CH–COOH$$
$$\downarrow +H_2O$$
$$R–CH(OH)–CH_2–COOH\ (\beta\text{-hydroxy acid})$$
$$\downarrow -H_2$$
$$R–CO–CH_2–COOH\ (\beta\text{-keto acid})$$
$$\downarrow -Co_2$$
$$R–CO–CH_3\ (\text{Methyl ketone})$$

Methyl ketone is also produced from keto acid naturally present in milk fat during heat treatment and storage.

References

Fox, P. E. and McSweeney, P. L. H. *Dairy Chemistry and Biochemistry*. Madras: Blackie Academic & Professional, 1998.

http://ecoursesonline.iasri.res.in/

Jenness, R. and Patton, S. *Principles of Dairy Chemistry*. New York: John Wiley & Sons, Inc., 1959.

Mathur, M. P., Datta Roy, D., and Dinakar, P. *Text Book of Dairy Chemistry*. Delhi: Indian Council of Agricultural Research Govt. of India, 1999.

Webb, B. H., Johnson, A., and Alford, J.A. *Fundamentals of Dairy Chemistry*. West Port, Connecticut: The AVI Publ. Co. Inc., 1978.

5

Milk Salt

The salts of the milk include those constituents that are present as ions or in equilibrium with ions. The main salts of milk are phosphates, citrates, chlorides, sulfates, carbonates, and bicarbonates of sodium, potassium, calcium, and magnesium. Approximately 20 other elements are found in milk in trace quantities including copper, iron, lead, boron, manganese, zinc, iodine, etc. These ions are more or less associated with themselves and with proteins. Some metals act as catalysts in the oxidation of milk fat leading to oxidative rancidity and certain metals like Ca, P, Mg, Mo, Co, Fe, etc. are components of metalloproteins such as lactoferrin, lactoperoxidase, xanthine oxidase, and parts of phospholipids.

The minerals in human, cow, and buffalo milk play an important role in the availability of these minerals for the young ones. They also play an important role in the digestibility of milk proteins. Na, K, and Cl are primarily present as free ions but Ca, Mg, phosphate, and citrate are distributed throughout many complexes (the highest concentrations are $CaCit^-$, Mg Cit^-, $H_2PO_4^-$, HPO_4^-, and $CaHPO_4^-$) (Table 1).

Table 1: Important salts in bovine milk.

	Concentration of salts	
Constituents	**Mean (mg/100 g)**	**Range (mg/100 g)**
Cationic		
Sodium	58	47–77
Potassium	140	113–171
Calcium	118	111–120
Magnesium	12	11–13
Anionic		
Total Phosphorus	74	61–79
Inorganic phosphorus	63	52–70
Ester phosphorus	11	8–13
Chloride	104	90–127

Sulphate	10	-
Citrate as citric acid	175	-

Source: Miscellaneous.

Trace Elements

The large number of trace elements found in milk have a significant influence on the various properties of milk and a significant role in human nutrition. The concentration of elements like I, Mo, and Zn in the milk exhibits large variations and depends markedly on the diet consumed by the cow. The concentrations of some of them are increased by contamination with utensils and equipment to which milk is exposed while handling and processing (Table 2).

Table 2: The concentration of trace elements in milk.

	Concentration (µg/L)	
Element	**Range**	**Typical value**
Aluminium (Al)	150–1000	500
Arsenic (As)	30–60	-
Barium (Ba)	-	Trace
Boron(B)	100–1000	300
Bromine (Br)	500–20000	-
Cadmium (Cd)	1–30	-
Cesium (Cs)	-	Trace
Chromium (Cr)	5–80	15
Cobalt (Co)	0.4–1.0	0.5
Copper (Cu)	10–200	-
Fluorine (F)	70–220	-
Iodine (I)	10–1000	-
Iron (Fe)	100–1500	300
Lead (Pb)	20–80	40
Lithium (Li)	-	Trace
Manganese (Mn)	20–100	50
Mercury (Hg)	-	Trace
Molybdenum (Mo)	20–120	70
Nickel (Ni)	0.30–0.8	-
Rubidium (Rb)	100–3400	-
Selenium (Se)	4–1200	12
Silicon (Si)	-	1400
Silver (Ag)	15–50	45
Strontium (Sr)	40–500	170

Tin (Sn)	-	Trace
Titanium (Ti)	-	Trace
Vanadium (V)	-	Trace
Zinc (Zn)	2000–5000	3300

Source: Miscellaneous.

Distribution of Salts Between Casein Micelles and Serum

The phosphorus is present as orthophosphate, but part of it is bound to organic components like serine and threonine residues of casein, molecules of hexoses and glycerol, phospholipids, etc. Phosphorus is also found in the hydroxyapatite structure of the colloidal calcium phosphate in the micellar structure of casein. The sulfur content of milk is about 0.36 g/kg, but most of it is in the amino acid residues methionine and cysteine of the proteins. About 10% is present as inorganic sulfate.

Physical Equilibria Among Salts

Milk contains several elements but all of them are not entirely in a soluble state. Some minerals exist in colloidal and ionic states at the normal pH of milk. The compounds that are in soluble condition play an important role in keeping various milk constituents in stable condition. A balance exists between the components that are in a soluble state and those which are in colloidal state.

Salt Solution in Milk

- The dissolved salts of milk are phosphate, citrate, chloride, sulfate, bicarbonate, sodium, potassium, magnesium, and calcium.
- Physical equilibria among the milk salts for the stability of milk is necessary, especially during heat processing. Colloidal particles of casein contain predominantly calcium, magnesium, phosphate, and citrate. The chloride and sulfate are entirely present as the free ions Cl^- and SO_4^{2-} at the pH of milk.
- The salts of the various weak acids (phosphates, citrates, and carbonates) are distributed among various ionic forms. Calcium and magnesium form complex soluble ions with citrate, phosphate ($CaPO_4$), and bicarbonates ($CaHCO_3^+$). The formation of such complex ions has the effect of reducing the concentration of calcium and magnesium ions in the solution.

Salt Balance in Milk

- Salts in milk exist in colloidal and soluble form.
- A large amount of salt is present in the casein micelle structure and is in equilibrium with salts in the soluble phase of milk.
- During the acidification/fermentation of milk, some salts get solubilized from the colloidal state and come into the soluble state. This destabilizes the casein micelle structure and tends to precipitate out, as happens during the curd setting of milk and the preparation of acid casein.
- It also happens that the casein micelle stability is lost during the heating of milk at high temperatures, such as in UHT processing, sterilization, and boiling of milk. This is due to the solubilization of colloidal calcium phosphate and the final loss in the stability of milk, this in turn affects the quality of dairy products. Such changes in the salt balance also take place during the manufacture of condensed milk, evaporated milk, and milk powder.

Salt Balance in Milk

The balance between certain salts such as Ca^{+2} and Mg^{+2} with citrate and phosphate in the colloidal and soluble phase plays a pivotal role in the stability of casein micelles structure.

The salt balance in milk is defined by the following equation

$$\text{Salt balance} = \frac{Ca^{+2} + Mg^{-2}}{\text{Citrate}^{-3} + PO_4^{-3}}$$

A delicate balance of cations and anions in the colloidal and soluble state is important in determining the stability of milk. The monovalent cations (Na^{+}) have a dispersing effect, while divalent and trivalent cations (Ca^{+2} and Mg^{+2}) have an aggregating effect. Therefore, the addition of sodium citrates and phosphate improves the heat stability of milk.

Colloidal salts remain in equilibrium with the dissolved salts. Various treatments of milk may cause the transfer of salts from one phase to the next. Two-thirds of the calcium, one-third of the magnesium, and half of the phosphorus are colloidal in normal milk.

It is presumed that all the colloidal phosphorus is present in the micellar calcium phosphate. A small portion of colloidal calcium is bound to α-lactalbumin. β-lactoglobulin can also bind calcium and magnesium and there are other minor calcium-binding proteins. Nearly all the sodium, potassium, and chlorine are diffusible.

The principle dissolved salt constituents of milk consist of phosphate, citrate, chloride, sulfate, bicarbonate, sodium, potassium, magnesium, and calcium. At pH 6.6 of milk, sodium and potassium are not found in any combination with other constituents and are present as the cations Na^+ and K^+ while; the chloride and sulfate (salts of strong acids) are present as free ion Cl and SO_4^-. However, the salts of the weak acids (phosphate, citrates, and carbonates) are distributed among various ionic forms. Calcium and magnesium form soluble complex ions with citrate, phosphate, and bicarbonates.

Salt Balance Theory of Sommer and Hart

Sommer and Hart were the first to demonstrate that aside from albumin, salt balance has a greater effect on heat stability, and salt balance is readily changed by other changes like acidity, thus affecting coagulation temperature.

- It was observed that casein has maximum heat stability when in combination with a definite optimum amount of calcium. When the Ca^{++} content available for the calcium casein complex is above or below this optimum combination, the casein is less stable to heat. The calcium contained in the milk distributes itself between casein, phosphates, and citrates. In addition, the Mg present reacts by replacing the Ca in the PO_4 and citrates. The effect of calcium and Mg being basic radicals is opposed to the effect of PO_4 and citrates which are acid radicals. The calcium casein combination is at its optimum heat stability when the above two groups of mineral salts are in balance, hence the term salt balance. An excess or deficiency of either group accelerates heat coagulation.
- If coagulation in heat test is due to a deficiency of Ca and Mg, it can be prevented by the addition of a proper amount of soluble Ca or Mg salts such as Ca and Mg acetates or chlorides. Such milk may also be stabilized by a slight increase in acidity because the increased acidity changes secondary PO_4 to primary PO_4 and the primary PO_4 has little or no effect on the salt balance. This change diminishes the amount of PO_4 that ties up the Ca and more Ca^{++} is available to satisfy the Ca^{++}—casein equilibrium. In such cases, a slight increase in acid thus improves the Ca-casein balance and raises the heat coagulation point.
- If troublesome heat coagulation is due to high Ca^{++} and Mg^{++} it can be prevented by the addition of the proper amount of PO_4^- or citrate such as di-sodium phosphate or sodium citrate.
- It is found in most cases that low heat stability is due to excess of Ca and Mg. Hence the heat stability is mainly controlled by the exclusive addition of salts of PO_4 and citrates. And since PO_4 is the cheaper of the two, it is preferred over citrate.

- The variation in the ion concentration such as buffer salts such as PO_4, and citrate especially those of citrate have a greater effect on the coagulation temperature than a slight variation in the pH of normal milk within the range of 6.58–6.60. But both factors must be considered.
- It is concluded that each lot of milk represents a separate colloidal system and that for each system there is an optimum combination of salt balance, due to such factors as pH with which optimum heat stability is attained.
- Thus, the main effect of salt composition is through the calcium and phosphate contents. The addition of a certain salt to milk can strongly disturb all salt equilibriums involved. The addition of calcium and phosphate to milk, up to the concentrations that are found in concentrated milk, causes its heat stability at $pH > 6.8$ to be equal to that of concentrated milk, i.e. zero.

Factors of Salt Equilibria in Milk

Temperature

The temperature will shift the balance among the various forms as milk is subjected to various cooling and heating treatments after it is drawn from the cow at 37°C. With the rise in the temperature the dissolved will be transferred to the colloidal phase. Similarly, lowering the temperature below that at which the milk is drawn would cause a transfer of calcium and phosphate from the colloidal particles to the dissolved state.

Acidity

There will be a pronounced shift in the salt equilibria with the addition of acid to milk whether directly or indirectly by bacterial action. A decrease in the pH withdraws the calcium and phosphate from colloidal particles until at about pH 5.2 all the calcium and phosphate are in the dissolved state.

Carbon dioxide Content

Milk as secreted by the cow contains about 20 mg of CO_2 from milk is accelerated by heating and agitation. Removal of CO_2 would affect the balance in the rest of the system. It would be expected that the removal of CO_2 and the consequent rise in pH would be reflected in a shift in calcium phosphate from the dissolved to the colloidal state and probably also in a shift of calcium ions activity per 100 mL or about 10% by volume.

Concentration of Milk

Concentrated milk tends calcium phosphate and calcium citrate to accumulate in the colloidal particles because the solubility is exceeded. As these materials are insolubilized, hydrogen ions are liberated, lowering the pH. The net result is an increase in the concentration of citrate and phosphate in both the dissolved and colloidal states.

Sequestering Agents and Ion Exchangers

To stabilize the milk or even to improve the utility of milk for a particular purpose it is desirable to treat milk to alter its ionic balance. The best-known example of this is the addition of phosphate or citrate to stabilize milk against subsequent heat coagulation. The addition of phosphate or citrate results in the binding of more of the calcium in the form of soluble complexes and decreases the activity of the calcium ions.

References

Fox, P.E. and McSweeney, P.L.H. *Dairy Chemistry and Biochemistry*. Madras: Blackie Academic & Professional, 1998.

Jenness, R. and Patton, S. *Principles of Dairy Chemistry*. New York: John Wiley & Sons, Inc., 1959.

Mathur, M. P., Datta Roy, D., and Dinakar, P. *Text Book of Dairy Chemistry*. Delhi: Indian Council of Agricultural Research. Govt. of India, 1999.

Webb, B. H., Johnson, A., and Alford, J. A. 1978. *Fundamentals of Dairy Chemistry*. West Port, Connecticut: The AVI Publ. Co. Inc.

6

Milk Enzyme

There are Two Types of Enzymes in Milk

1. Endogenous and 2. Exogenous.

Three sources of enzyme in milk:

1. The blood via defective mammary cells.
2. Secretory cell cytoplasm. Some of which are occasionally entrapped with fat globules by the encircling MFGM.
3. The outer layer of MFGM is probably the principal source of endogenous enzymes in milk. The outer layer of MFGM is derived from the apical membrane of the secretory cell which in turn originates from the Golgi membrane and acts as the principal source of enzyme in milk.

Almost 60 endogenous enzymes have been reported in bovine milk.

Table 1: Milk enzymes and their importance

Enzymes	Importance
1. Lipase	• Off flavors in milk. • Flavor development in blue cheese.
2. Proteinase	• Formation of ϒ-Cn from β-Cn during cheese ripening. • Reduces storage stability of UHT milk.
3. Catalase	• Indicator of mastitis. • It acts as a pro-oxidant.
4. Lysozyme	• Antimicrobial agent.
5. Xanthine oxidase	• Helps cheese ripening • Prooxidant
6. Sulfhydryl oxidase	• Antioxidant • Responsible for cooked flavor
7. Lactoperoxidase	• Index of pasteurization. • Antimicrobial agent. • Index of mastitis. • Prooxidant.
8. Alkaline phosphatase	• Index of pasteurization.
9. Acid phosphatase	• Reduces heat stability of milk. • Helps in ripening.

General Characteristics of Milk Enzymes

1. Deteriorative character e.g. lipase (commercially most important enzyme in milk), proteinase, acid phosphatase, xanthine oxidase.
2. Preservative enzymes. e.g. superoxide dismutase, sulfhydryl oxidase.
3. As an indicator of the thermal treatment history of milk. Related enzymes are alkaline phosphatase, ϒ glutamyl transpeptidase, and lactoperoxidase.
4. Mastitis infection indicator. Related enzymes are catalase, n-Acetyl-β-D-glucosaminidase, and acid phosphatase.
5. Antimicrobial activity.
6. Commercial source of ribonuclease and lactoperoxidase.

Common Milk Enzymes and their Role in Dairy Industry

Lipase

Lipase catalyzes the development of hydrolytic rancidity in milk and milk products.

Classification

- A-type of carboxylic ester hydroxylases or arylesterases.
- B B-type esterases e.g. glyceryl tricarboxylic esterases, aliphatic esterases.
- C-type esterases e.g. choline esterases.

Role of A-type Esterase

- It hydrolyzes aromatic esters e.g. phenyl acetate.
- It shows little activity on tributyrin and is not inhibited by organophosphate.

Role of B-type Esterase

- B-type esterase is the most active among aliphatic esters.
- Sometimes they show their activity on aromatic ester.
- Organo phosphate compounds inhibit this enzyme

Role of C-type Esterase

- They are most active on choline ester.
- They can also hydrolyze some aromatic and aliphatic esters very slowly.
- They are inhibited by organophosphate.

Lipases in Milk

In normal milk the ratio of A-type, B-type, and C-type is 3:10:1. Mastitis infection increases A esterase activity considerably.

A and C esterases have little technological significance in milk.

Mechanism of Lipase

- Lipase hydrolyzes ester bond in emulsified ester i.e. water–oil interface.
- Blood serum albumin and calcium increase the activity that binds free fatty acids (optimal activity of lipase is observed at pH 9.2 at 37° C).
- LPL activated by lipo-protein co-factor is the principle indigenous milk lipase which attracts special focus.
- The lipolytic system in the milk becomes active during the damage of MFGM by agitation, homogenization, or temperature fluctuation.
- Some individual cows produce milk which shows rancidity development spontaneously. Membrane lipase is considered to be the main reason behind this.

Occurrence

Casein micelle Association of lipase is more than 90% while the triglyceride substrates are in fat globules surrounded and protected by fat globule membrane.

Significance of Lipase

Technologically lipase is considered to be the most significant enzyme in milk.

Indigenous milk lipase plays a significant role in cheese ripening but the most industrially important aspect of milk lipase is its role in inducing hydrolytic rancidity which makes milk and milk products unpalatable. All milk contains an adequate level of lipase but the lipase activity is demonstrated only after the fat globule membrane has been damaged.

Xanthine Oxidase

Isolation

Xanthine oxidase is concentrated in MFGM and is considered a principal protein of MFGM. Therefore, the isolation of xanthine oxidase mainly involves cream as a starting material. A dissociating agent generally liberates xanthine oxidase from membrane lipoprotein.

Properties

1. Milk xanthine oxidase has a molecular weight of 300 kDa.
2. It consists of two subunits.

3. The optimum pH of xanthine oxidase is 8.5.
4. The enzyme requires Flavin adenine dinucleotide (FAD), Fe, Mo, and acid labile compounds as cofactors (animals which are having low Mo exhibit lower xanthine oxidase activity).
5. Xanthine oxidase has five genetic variants.

Xanthine Oxidase Action in Milk

1. Various processing operations affect xanthine oxidase action in milk.
2. Xanthine oxidase activity is enhanced during storage at 4°C/24 h. The increase is almost 100%. Homogenization also increases the activity of xanthine oxidase by 60–80%.
3. Heating milk at 73°C/5 min also increases the activity of xanthine oxidase by 50–100%.

    These treatments (homogenization/freezing) cause the transfer of xanthine oxidase from the fat phase to the aqueous phase. The stability of xanthine oxidase on heating is dependent on the phase of the enzyme. Aging and homogenization increase the susceptibility of xanthine oxidase to heat. Xanthine oxidase in cream is most heat stable and the lowest stability to heat is observed in skim milk. Homogenization releases active undenatured xanthine oxidase from the milk fat globule membrane.

Significance in Milk and Milk Products

Xanthine oxidase is important mainly from the two aspects: lipid oxidation and atherosclerosis.

- Lipid oxidation: xanthine oxidase is a prooxidant. It excites stable triplet oxygen.
- Milk that undergoes spontaneous rancidity contains about 10 times more xanthine oxidase than normal milk.
- Spontaneous oxidation is generally induced in milk by adding xanthine oxidase, four times than normal level present in milk.
- Heat-denatured/flavin-free xanthine oxidase is inactive.
- Susceptibility of unsaturated fatty acid to oxidation caused by xanthine oxidase increases with the degree of unsaturation.

Atherosclerosis: Xanthine oxidase may cause the narrowing of an artery. Xanthine oxidase from homogenized milk enters the vascular system and induces atherosclerosis by the lipid oxidation of the cell membrane plasminogen in the cell membrane. This oxidation of plasminogen may cause atherosclerosis.

Catalase

Catalase is a pro-oxidant. Its activity is observed in the skim milk phase.

Isolation

Catalase is purified from the pallet obtained from buttermilk by centrifugation at 10,000*g*.

Characteristics

- Milk catalase is a heme protein having a molecular weight of 200 kDa.
- The isoelectric pH is 5.5.
- Catalase is more stable at pH 5–10 but rapidly loses its activity outside this range.

Inactivation

Complete inactivation of catalase may be done by heating the milk at 71°C for 1 h.

Hg, Fe, and Cu also inactivate catalase.

Activity

- Feed and stage of lactation regulate the activity of catalase in milk.
- Mastitic infection affects milk catalase activity. So, the activity of catalase in milk may act as an index for detecting mastitis infection.
- Acts as a lipid prooxidant due to the presence of iron-containing heme iron present in it.

Lysozyme

Source

Lysozyme is generally isolated mainly from human and equine milk. Lysozyme in human milk has special importance due to its high antibacterial activity. It is very important for human nutrition.

Baby food formula uses supplementation of lysozyme with egg white lysozyme.

Role of Lysozyme

Lysozyme is a very good antibacterial agent. It inhibits the growth of vegetative cells of *Clostridium tyrobutyricum* and hinders the germination of its spores. Lysozyme also kills *Listeria monocytogenes*.

Uses

Lysozyme addition permits using of lower temperatures in food sterilization.

Pro-immobilized lysozyme has been proposed for self-sanitizing in the mobilized enzyme column.

Lactoperoxidase

Lactoperoxidase is considered to be the most heat-stable enzyme in milk and its destruction was used as an index of flash pasteurization. It is also used as an index of super HTST pasteurization. Lactoperoxidase was first isolated in 1943.

Chemical Properties

Lactoperoxidase is a heme protein containing about 0.7% iron. The optimum pH of Lactoperoxidase is 3. The molecular weight is 77.5 kDa. It consists of two identical sub-units. It has 10 variants which are dependent on its amyl group (glutamyl and asparagine) and carbohydrate group.

Significance

LP is technologically significant for various reasons.

1. It may be considered as an index of mastitic infection. Lactoperoxide levels increase during mastitic infection.
2. Lactoperoxidase causes nonenzymatic oxidation of unsaturated fat. This is due to the protein of the heme group present in its structure. Heat-denatured enzymes are more active than native enzymes.
3. Lactoperoxidase in combination with hydrogen peroxide and thiocyanate acts as a potential antimicrobial agent. Hydrogen peroxide is added from outside of the system. (thiocyanate is produced from *Brassica* sp.)
4. Acid production in milk by some starter microorganisms is reported to be retarded by severe heat treatment. Temperature 77–80°C/10 min. This system can be restored by the addition of lactoperoxidase.

Galactosidase Transferase (GT)

It catalyzes the transfer of galactose.

Role of GT

Disaccharide, oligosaccharide, and polysaccharide biosynthesis in milk involves galactosyl tranferase activity. This enzyme catalyzes the transfer of sugar moiety from an activated molecule to a specific sugar molecule forming a glycosidic bond.

GT family comprises enzymes with several known activities. These are N-acetyl lactosaminide β-1,3-N-acetyl glucosaminyl transferase.

Role in Milk

GT plays a significant role in the biosynthesis of lactose. Lactose is synthesized from glucose absorbed from the blood. UDP-galactose, an intermediate component of lactose synthesis is linked to glucose by a reaction catalyzed by an enzyme called lactose synthetase. GT is a major component of lactose synthetase. It transfers the galactose from UDP galactose to a no of accepting.

Phosphatase

There are two types of phosphatase namely: alkaline phosphatase/alkaline phosphomonoesterase and acid phosphatase/acid phosphomonoesterase.

Another phosphatase is also present in milk but doesn't exhibit any function.

Alkaline phosphatase/alkaline phosphomonoesterase:

Characteristics of Alkaline Phosphatase

The optimum pH is 6.8. The optimum temperature for activity is 37°C. Molecular weight: 170–190 kDa, number of polymeric forms—4.

Activators: calcium, magnesium, zinc, cobalt, and manganese.

Role of Alkaline Phosphatase

Alkaline phosphatase is milk-significant because the time and temperature combination required for the inactivation of alkaline phosphates were slightly more severe than those required to inactivate Mycobacterium tuberculosis (the target microorganism for pasteurization).

The phosphatase test to assess pasteurization efficiency is based on this property of alkaline phosphatase. The substrates used for the phosphatase test are phenyl phosphatase, and para-nitrophenyl phosphate/phenolphthalein phosphate.

Presence in Milk

Alkaline phosphatase is concentrated on MFGM and hence in cream. It is released into the buttermilk during phase inversion.

Acid Phosphatase/Acid Phosphomonoesterase

Characteristics

1. Milk contains acid phosphatase which has an optimum pH 4.
2. It is very heat stable.

3. Denaturation of acid phosphatase under UHT reaction follows 1st-order kinetics.
4. This enzyme retains significant activity following HTST pasteurization but does not survive in bottle sterilization or UHT treatment.
5. This enzyme is not activated by magnesium. It is inhibited by chlorine.
6. The molecular weight of acid phosphatase is about 42 kDa.

Significance

Acid phosphatase is present in milk in much lower concentration than alkaline phosphates but it has higher heat stability and lower pH optimum. The acid phosphatase activity in milk increases 4–10 times during mastitic infection.

Is assumed that indigenous/bacterial activity phosphatase is mainly responsible for dephosphorylation of cheese.

References

Fox, P. E. and McSweeney, P. L. H. Dairy Chemistry and Biochemistry. Madras: Blackie Academic & Professional, 1998.

Jenness, R. and Patton, S. Principles of Dairy Chemistry. : John Wiley & Sons, Inc., 1959. Mathur, M. P., Datta Roy, D., and Dinakar, P. Text Book of Dairy Chemistry. Delhi: Indian Council of Agricultural Research. Govt. of India, 1999.

Webb, B. H., Johnson, A., and Alford, J. A. Fundamentals of Dairy Chemistry. West Port, Connecticut: The AVI Publ. Co. Inc., 1978.

7

Chemistry of Cream and Butter

Chemistry of Creaming

After being drawn from animals, normal milk will form a cream layer on standing. Because of the difference in density between fat globules (920 kg m^{-3} at room temperature) and milk plasma (density 1030 kg m^{-3}), the granules tend to rise. This causes creaming, which we often want to prevent and it enables milk to be separated into cream and skim milk. Creaming is much enhanced when the globules have been aggregated into flocculates, clusters, or granules.

Theory of Creaming

From Archimedes' principle—the field force on a submersed spherical particle of diameter d is $\frac{1}{6} a\pi d^3 (p_p - p_p)$

where p_p is the density of the plasma, p_f is the density of fat, a is the acceleration defining the field whether from gravity or centrifugation. The globules attain a velocity of v and according to Stoke's the frictional force acting on the globule is given by

$3\pi dn_{pv}$,

where n_p is the viscosity of the plasma, by gathering, both forces equally, we obtain Stoke velocity

$v = 1/18\, a \cdot d^2 \cdot (p_p p_f)/n_p$

For gravity creaming an equals g (9.81 ms^2)

In a centrifugal field, $a = Rw^2$,

where R is the effective radius of the centrifuge, w is the angular velocity in radians per second.

In a milk separator, a is usually about $4000 \times g$.

Stokes equation is applicable under the following conditions:

1. The concentration of fat globules must be very low.
2. Creaming may not be disturbed by Brownian motion or convection currents. Disturbance always occurs for small fat globules (<1 μm) during gravity creaming.

3. Globules should be smooth spheres. This is not true for most homogenized fat globules.

Cream Ripening

This unit operation serves as an intensive preparation of the cream for the churning process. It is divided into two categories.

1. Physical ripening;
2. biochemical ripening

The following criteria are influenced by the cream ripening:

- Consistency, firmness, and spreadability of butter.
- Basic water content of butter during the manufacturing process (bound water in butter, which cannot be removed by any mechanical treatment).
- Buttermilk fat content, which has to be minimized.
- Acid content and flavor of the butter during the biochemical ripening.

Physical ripening is required irrespective of the type of butter manufactured.

For the manufacture of sour cream butter, the biochemical ripening which causes the acidification of the cream is combined with the physical ripening.

Physical Ripening

Physical ripening serves to restructure the liquid fat (after heating) into a more or less crystalline state, which is the basic requirement for a solid, plastified structure in butter.

1. Physical ripening is influenced by the following factors:
2. Fat composition,melting and solidification performance,
3. Melting and solidification curves,
4. Influence of the fat globule membrane.

Fat Composition

The melting and solidification behavior of fat depends on the attached fatty acid, which indicate the hardness of the butterfat.

Natural hardness is caused mainly by seasonal feeding of the milk animals.

Winter feed with a great deal of dry feed or roughage yields a hard butterfat with a high percentage of saturated and short-chain fatty acids and few double bonds in the fat molecules.

Summer feeding (green grass) results in fat with longer chains and a higher percentage of unsaturated fatty acids with many double bonds.

We therefore distinguish between summer fat and winter fat or summer cream and winter cream.

The percentage of saturated and unsaturated fatty acids can be determined from the iodine number or the refractive index, from which we can estimate the hardness and sometimes the solidification performance of the fat.

The iodine number indicates the amount of iodine (halogen), which is required to saturate the double bonding in 100 g of fat.

If the refractive index RI is determined with a butter refractometer, the iodine number can be calculated as follows:

$I = 3.81 \times \text{RI} - 128.85$

Melting and Solidification Performance

The type and means of crystallization (crystallization kinetics) can be influenced by the temperature profile of the cream. When the fat exists between the melting and solidification temperature, each further temperature reduction disturbs the solution equilibrium of the fat molecules, which results in a supersaturated state of the solution. As a results initially small crystals are formed, which grow steadily in to larger crystal structure. Fat crystallization is a slow process, which can last for hours, even for days.

Four different fat globule types can occur at a constant temperature profile of 11–15°C.

Type 1: Fat globules with thin peripheral crystal layer and a liquid interior.

Type 2: Fat globules with thin peripheral crystal layer along with crystalline fat agglomerates and a little liquid fat in the interior.

Type 3: Fat globules with thick crystalline shell and a liquid core.

Type 4: Fat globules with thick crystalline shell along with crystalline agglomerates and a little liquid fat in the interior.The frequency of the occurrence of these types can be controlled by the variation in the temperature profile. This makes it possible to influence the butter consistency relatively independent of the fat composition.

Melting and Solidification Curves

The determination of iodine number is insufficient to achieve optimal solidification by using a controlled temperature profile because iodine number characterizes only the ratio of hard (or higher melting) to soft (lower melting) fats. It becomes necessary to record the melting and solidification curves, from which the peaks of the temperature set points during the tempering of the cream can be derived (Table 1).

Influence of the Fat Globule Membrane

During fat crystallization, there is a reduction in the volume of the fat globule. This causes a reduction of the boundary layer forces in the area of the membrane toward the aqueous phase, and the fat globule membranes become unstable through the peripheral crystalline layer. Further, the crystals damage the membranes to such an extent that the liquid fat can migrate through the fat globule membrane during mechanical treatment.

Table 1: Optimal crystallization temperature of butterfat as a function of iodine number and refractive index.

Iodine number	Refractive index	Optimal crystallization temperature
<28	<41.2	6–7°C
29–31	41.3–42.0	
32–34	42.1–42.8	
35–37	42.9–43.6	>10°C
38–40	43.7–44.3	
>40	>44.4	

Ripening Process

The ripening process is basically a time temperature process, to which the cream is subjected to. There are four types of ripening process namely,

1. Cold/warm/cold ripening (Alnarp process).
2. Warm/cold/cold ripening.
3. Warm ripening.
4. Cold ripening.

The cold/warm/cold ripening process is used during the winter months for hard butter fat, as here we see mainly the formation of type 4 fat globules with a thick crystalline shell. The result is that during the buttering process, a high percentage of fat globules remain intact, and the butter has low hardness (0.8 N), i.e. it remains soft.

The warm/cold/cold ripening is mainly suited for soft fat (with typical summer feeding) and leads to fat globules without shells of type 2, which are easily destroyed during the buttering. This butter shows only a few intact fat globules with a relatively homogeneous distribution of crystalline fat. The hardness of such butter is significantly higher (1.5 N) than the butter obtained from the cold/warm/cold process.

Butter Churning

In the buttering machine, strong mechanical forces act on the cream. Fat globule membrane parts are removed by the higher shear stress until such time that the fat crystals penetrate and destroy the membrane. The fat globules are pushed together simultaneously in the space between the beater and the wall of the buttering machine; a discharge of liquid fat passes through the membrane, and the fat globules agglomerate. Solid fat, intact fat globules, and serum droplets are entrained, and because of the tumbling movement of the liquid, a grainy product with nearly the same grain size is formed. These pellets consist of many small, fat grains and are called butter grains.

Factors of Churning

Butter formation is influenced by

1. The quality of the cream and
2. The buttering machine.

The following parameters of cream influence butter formation:

I. Fat content: higher fat content leads to quick butter formation.

II. Buttering temperature: higher temperature leads to quick butter formation.

III. Ripening degree; optimal p^H.

In the buttering machine, butter formations happen within 5–15 s. The buttering time is influenced by factors such as:

i) Rotational speed of the beaters, variable with the shaft speed (low speed longer buttering time).

ii) Cylinder length and diameter.

iii) Radial distance between beater and cylinder.

iv) Cream flow rate (high flow rate results in short buttering time).

Biochemical Ripening

During the physical ripening time–temperature processes are incorporated, which generate biochemical reactions when lactic acid-producing cultures are added. The objective is:

- To generate a typical acid level (P^H value) and flavor, which characterize sour cream butter.

When cultures are added (containing the lactic acid bacteria), lactose is fermented through various steps into pyruvic acid, which is then reduced to lactic acid.

$$\underset{\text{Pyruvic acid}}{\text{CH–CO–COOH}} + H_2 \longrightarrow \underset{\text{lactic acid}}{\text{CH--CH (OH)–COOH}}$$

There are also trace amounts of propionic acid, acetic acid, and carbon dioxide, among other fermentation products. The generated acid causes some casein denaturation, which modifies the dispersed cream system in such a way that the fat can be released more easily from the serum during butter formation.

Further formation of the diacetyl flavor of butter occurs at a p^H of 5.0–5.2; lactic acid displaces citric acid from citric acid salt, and the released free citric acid in the presence of lactose is converted into acetone and diacetyl by the flavor compounds. The same fermentation products can form flavor precursors from the citrates directly.

Diacetyl formation can be enhanced by

- Incorporation of air by intensive stirring of the cream during the initial part of the ripening time.
- Ripening temperature <15 °C in the second part of the ripening period.
- Maintaining the optimal pH (<5.2).
- Addition of 0.15% citric acid to the cream.

The diacetyl amount in sour cream butter is 0.5–2.0 mg/kg butter.

The correct ripening stage for buttering has been reached by the cream when there is sufficient fat crystallization and a proper p^H value. According to the type of butter and the quantity, the p^H is found in the range of 4.8–5.2. The proper ripening stage can also be found by determining the percentage of lactic acid developed (titratable acidity).

Nutrients in Cream

The nutrients in the cream reflect those present in the unseparated milk. The degree of separation causes the levels of water-soluble nutrients to fall, while those of fat-soluble nutrients increase. Vitamin A content is higher in cream, which has a higher percentage of fat than whole milk. Losses during processing largely correspond to those of milk. Losses of vitamin C and folic acid are of particular significance in UHT and in-container sterilized cream. Significant losses of vitamin B_{12} also occur during in-container sterilization.

Flavor and Aroma of Cream

The characteristic flavor and aroma of cream are derived primarily from the fat constituents, although the contribution from constituents of the aqueous phase and the MFGM is also significant. Alkanoic acids ($C_{10,12}$), d-lactone, indole skatole, di-methyl di-sulfide and hydrogen sulfide, at the levels commonly present in cream, are considered to contribute to the desired flavor, with a marginal contribution from phenol and phenolic compounds such as O-methoxy phenol. Oxidation during whipping may improve flavor and 4-cis-heptanol, if present in micrograms per kilogram contributes to the full flavor.Lipolysis may occur during the production of double cream due to milk lipase activity in the separated cream before pasteurization and is enhanced by the rewarming of cold milk.

Deteriorative changes in cream flavor during storage result primarily from fat lipolysis or oxidation. Cream treated with insufficient heating to inactivate casein-associated lipase is, theoretically, highly prone to lipolysis, but in practice, the life of such cream is likely to be limited by bacterial spoilage. Lipolytic enzymes produced by psychrotrophic bacteria in raw milk are of greatest significance in UHT-sterilized milk. The spoilage potential is high in cream due to the high fat content and the tendency of psychrotroph lipases to partition into the fat phase. However, free fatty acids are less readily detectable in cream than milk, contributing to an overall perception of deterioration. Psychrotroph-derived protease may also cause spoilage involving thickening, gelation, and bitter flavor.Cream is highly prone to lipid oxidation, especially in light. Oxidation is accelerated by light in the wave length range 310–490 nm, but wavelengths of 440–490 nm are the most damaging lipid oxidation is rapid in raw and pasteurized cream, but free –SH groups formed from b-lactoglobuline during UHT or in-container sterilization provide a considerable degree of protection. Homogenized cream is more susceptible to oxidation as a greater surface area of fat is exposed to oxygen.

The milk fat globule membrane may play either a pro or antioxidant role, depending on other factors. The phospholipid oxidation in the membrane may trigger the triacylglycerol oxidation in fat globules, although it is also possible that metalloproteins such as xanthine oxidase are primarily responsible for oxidative capacity.

Composition and Standard of Butter

Butter is obtained from cream by the process of churning and working with the addition of salt. Butter is finally packed in wholesale containers or retail package. It generally has a pale yellow color, but varies from deep yellow to nearly white. Butter is a water-in-oil emulsion, which contains more than 80% milk fat, tiny droplets of water, milk SNF. It may contain salt also. The desired smoothness of butter comes from the working or kneading of the crystalline fat network during processing. Basically, butter contain butter fat, water and curd, the last consisting of casein, lactose, and mineral matter

Chemical composition (%) of butter

Fat	Moisture	Salt	Curd
80.2	16.3	2.5	1.0

According to FSSA (2006), butter means the fatty product derived exclusively from the milk of cows and/or buffaloes or its products, principally in the form of an emulsion of the type water-in-oil. The product may be with or without added common salt and starter cultures of harmless lactic acid and/or flavor-producing bacteria. The table butter shall be obtained from pasteurized milk and/or other milk products that have undergone adequate heat treatment to ensure microbial safety. It shall be free from animal body fat, vegetable oil and fat, mineral oil and added flavor. It shall have a pleasant taste and flavor free from off flavor and rancidity. It shall conform to the following parameter given in the table: where butter is sold or offered for sale without any indication as to whether it is a table or desi butter, the standards of the table butter shall apply (Table 2).

Table 2

Parameter	Table butter	Desi cooking butter
Moisture	Not more than 16.0% m/m	-
Milk Fat	Not less than 80.0% m/m	Not less than 76.0% m/m
Milk solids not fat	Not more than 1.5% m/m	-
Common salt	Not more than 3.0% m/m	-

Butter is usually classified into two main groups, mainly sweet cream, cultured, or sour cream butter. Butter can also be classified according to salt content: unsalted, salted, and extra salt.

Butter consists of unaltered fat globules and moisture droplets embedded in a continuous phase of butterfat. The structural properties are dependent on liquid to solid ratio in a continuous phase. The fatty flavor of the butter is due to diacetyl and to a lesser extent to butyric, acetic, propionic, and formic acids, acetaldehyde, and acetone. Butter produced from curd contains much more diacetyl than when the raw material is fresh cream but the amount can be increased by the addition of culture.

References

Bandyopadhyay, A. K., Ghatak, P.K., and Ray, P. R. Textbook on Analysis of Milk Products. New Delhi: Kalyani Publishers, 2016.

De, Sukumar. Outlines of Dairy Technology. New Delhi: Oxford University Press, 2004.

FSSA. The Food Safety and Standard Act. New Delhi: Professional Book Publishers, 2006.

Mathur, M. P., Datta Roy, D., and Dinakar, P. Text Book of Dairy Chemistry. New Delhi: Indian Council of Agricultural Research. Govt. of India, 1999.

Spreer, E. Milk and Dairy Product Technology. New York: Taylor & Francis, 2017.Varnam A. H. and Sutherland J. P. Milk and Milk Products: Technology, Chemistry and Microbiology. London: Chapman & Hall, 1994.

8

Chemistry of Cheese

Cheese is made from milk, with the addition of cream, buttermilk, and/or whey. This basic raw material is also called cheese milk. Cheese consists mainly of casein (the major protein of milk), other milk proteins, fat, other milk components, and a certain percentage of water, which is bound. Proteins are precipitated by enzymes and/or acidification as well as heat and are separated by suitable processes from the aqueous phase. (whey, serum). Cheese is most often molded, pressed, and salted and has fungal and bacterial culture additions. colorants, herbs, spices, and other non-dairy ingredients are also added to the cheese. It is consumed either fresh or at different stages of ripening.

Nutritional Fact

100g cheese has the following nutritional content (average):

Energy	1641 kJ (392 kcal)
Protein	23.7 g
Calcium	870 mg
Phosphorus	610 mg
Vitamin A	1740 IU
Vitamin D	13 IU
Riboflavin	0.50 mg
Vitamin B	0.0015 mg

Classification

Worldwide, there are more than 2000 types of cheese, sometimes made by very different manufacturing processes. A classification can be based on several aspects and is done in different countries according to different criteria. A general classification can be made for three major groups:

a. Rennet or natural cheese: manufactured straight from milk by using proteolytic enzymes (rennet) and acid, with a more or less pronounced ripening process.

b. Fresh cheese or non-ripened Cheese: Quarg, fresh cheese, and white cheese are examples. Its manufacturing process is similar to that of rennet cheese manufacturing, but it has a high degree of acidity and is not subjected to a proteolytic ripening process.

c. Long-life cheese (processed cheese): it is more often made from rennet cheese and is textured by thermal treatment and made shelf-stable.

For another classification into groups and types, different aspects and characteristics can be used, such as:

A. Type of process (rennet cheese, rennet acid cheese, acid curd cheese, etc.).
B. Type of consistency (hard, semi-soft, soft cheese, etc.).
C. Types of milk (Cow, ship, goat, buffalo, etc.).Chemical composition (Ca content in conjunction with P^H, dry matter, water fat).
D. Ripening process (ripened cheese and non-ripened fresh cheese).
E. Variations in taste.
F. Type of hole formation (large, medium small round holes, no holes).
G. Surface characteristics (blue fungus or white fungus cheese).

Rennet

Rennet is the preparation obtained commercially from the fourth or true stomach (abomasum) of the young calf, known as the VELL. Extraction of rennet is facilitated by using 15% NaCl brine.

The principal proteinase in such a rennet is chymosin. About 10% of the milk clotting activity of calf rennet is due to pepsin.

Like pepsin, chymosin is an a spartyl proteinase, i.e. it has two essential aspartyl residues in its active site, which is located in a cleft in the globular molecule (molecular mass approx. 36 kDa).

Rennet Types

Rennet is classified into the following categories:

1.	Animal rennet:	i. Calf rennet
		ii. Chicken pepsin
		iii. Bovine pepsin
		iv. Porcine pepsin
2.	Microbial rennet:	Acid proteinases from
		i. *Rhizomucor miehei*
		ii. *Rhizomucor pusillus*
		iii. *Cryophonectria parasitica*

3.	Vegetable rennet:	*Withania coagulans*
		Ficus spp.
		Papain

Cloned rennet

Cloning is the latest technique adopted for developing rennet-like enzymes. The gene responsible for producing calf chymosin is subjected to cloning in *Kluyveromyces marxianus var. lactis*, *Aspergillus niger*, and *E. coli.*

Microbial (cloned) chymosins have given excellent results during cheese-making trials on various varieties and are not widely used commercially. Significantly they are accepted for use in vegetarian cheese. The gene responsible for producing *Rhizomucor miehei* proteinase has been replicated in *A. oryzae*, the resultant product, Marzyme GM is commercially available.

Chemistry of Rennet Action

Rennet coagulation involves two distinct stages:

i. Primary enzymatic stage, a proteolytic stage where casein micelle is destabilized by hydrolysis of κ-casein to yield Para-κ-casein.

ii. Secondary non-enzymatic stage, a calcium-mediated stage in which Para-κ-casein micelles undergo limited aggregation.

i. **Primary enzymatic stage:** hydrolysis of κ-casein primarily involves cleavage of $Phe_{105} – Met_{106}$ linkage. This cleavage yields a Para-κ-casein (fl $_{105}$), common to all caseins, and a macro peptide unique to each κ-casein component.

Rennet

Casein →Para-κ-cascin + Macro peptides

↓ Ca^{++}, ~ 30 °C

Gel

Casein micelles are stabilized by k-casein, which represents 12–15% of total casein and is located mainly on the surface of the micelles in a way that its hydrophobic N-terminal reacts hydrophobically with the calcium sensitive a_{s1}, a_{s2}, and b caseins while its hydrophilic C-terminal region comes through the surrounding aqueous environment ‘stabilizing the micelles by a negative surface charge and steric stabilization.

Several physicochemical changes result directly from the hydrolysis of κ-casein

1. Liberation of the highly charged macro peptide C-terminal fraction of κ-casein from the micellar surface.
2. Reduction in the charge at the micelle surface (Zeta potential reduced from about 20 to 10 mv.
3. Decrease in micellar solvation, resulting from the decrease in inter micellar repulsion and casein water interaction, and a concurrent increase in inter micellar attraction.
4. Increased sensitivity of the Par-κ-casein micelles to aggregation.

ii. Secondary non-enzymatic stage: aggregation results from inter micellar cross-linking via calcium binding to serine-phosphate groups and commences when approximately 86% of total κ-casein has been hydrolyzed. Individual micelles, however, are only able to participate in aggregation when approximately 97% of their κ-casein has been hydrolyzed. This may happen due to residual intact κ-casein 'hairs' providing sufficient repulsion to prevent permanent contact between adjacent micelles and/or overcoming the attractive Van der Waals forces. Aggregation is, however, dependent on P^{H} value and temperature and occurs at a lesser degree of κ-casein hydrolysis at lower P^{H} values and higher temperatures.

Syneresis

Considerable contraction of rennet gel occurs on cutting, leading to expulsion of water from the curd. Syneresis rate is controlled by several factors, which are –

i. P^{H} value.
ii. Cooking temperature.

Three possible mechanisms of synersis have been proposed:

1. Changes in solubility.
2. Rearrangement of para-κ-casein network.
3. Shrinkage.

1. **Changes in solubility**: changes in solubility have formed the basis of mechanisms explaining shrinkage in polymer gels. Rennet curd is primarily a particle gel and while small regions around flocculated micelles may exist as a polymer gel, this is unlikely to be of significance. Changes in solubility may, however be of importance in acid-set curd.
2. **Rearrangement of para-k-casein network:** rearrangement of the network of Para casein particles provides a partial explanation of

changes during synersis of rennet curds. Rearrangement involves a more compact network formation with an increased number of bonds. It is challenging to achieve a more compact structure because of Par-k-casein's immobilization within the network. Despite this van der Waals and electrostatic attractive forces between flocculated micelles cause an increase in the number of bonds. There may also be limited cross-linking resulting from the thermal motion of the gel strands.

Each of the above mechanisms would increase tensile stresses on the gel strands, potentially leading to breakage and formation of new bonds. Breakage of strands could also result from external pressures, and re-arrangement likely becomes of increasing importance as synersis proceeds.

3. **Shrinkage:** Shrinkage of Para casein particles and thus of the gel is not thought to be an important mechanism under normal circumstances. Significant shrinkage does, however, occur if the P^{H} value is lowered, or the temperature raised.

Factors of Rennet Coagulation

The coagulation of milk is influenced mainly by the following factors:

1. Temperature
2. pH
3. Properties and concentration of proteins (substrate properties).
4 Concentration of rennet.
5. Calcium

Temperature

No coagulation occurs below 20°C, due mainly to the very high temperature co-efficient of the secondary phase. At higher temperature (above 55–60 °C), depending on P^{H} and enzyme, the rennet is denatured.

Rennet coagulation is prolonged or prevented by preheating milk at temperatures above 70ºC (depending on the length of exposure). The effect is due to the interaction of b-lactoblobuline with κ-casein via sulfhydryl—disulphide interchange reactions; both the primary and, especially, the secondary phase of coagulation is adversely affected.

pH

End of pre-ripening milk should have a cheese specific PH. This influences the enzymatic activity. Enzyme dosing is adjusted accordingly and the coagulation time is fixed.

Coagulation time is selected mostly as a function of technology and economy, and the enzyme quantity is closed accordingly. The rate of clotting increases rapidly with small increases in acidity. Alkalis considerably retard the clotting of milk by rennet.

Properties and Concentration of Protein

A key factor is the concentration of casein, which is the substrate for an enzymatic reaction. An additional factor is the salt equilibrium in milk, especially the calcium content and calcium solubility.

Realising substrate standardization is among the hardest things to do. The coagulation kinetics are influenced decisively by the protein content. Coagulation time can be adjusted to the quantity of enzymes utilized by changing the protein composition.

Concentration of Rennet

Rennet strength indicates the number of parts of cheese milk with an SH of about 7 and a temperature of 35 °C that can be curdled by one part of rennet in 40 min.

Calf rennet is offered in powdered form in a strength of 1:100000 or as a liquid rennet extract in a strength of 1:10000 or 15000. Rennilase is offered in a strength of 1:46000. Rennet strength can be calculated as follows:

$$L = V_M.100.2400S/V_L.T$$

L = Rennet strength, V_M = milk quantity in ml, V_L = Rennet solution used in ml, T = coagulation time in seconds, S = time unit.

Calcium

Calcium has little effect on the enzymatic stage of rennin action, while the second stage (coagulation of milk) is very sensitive to changes in concentration of calcium ions. It is common practice to add calcium chloride to milk which has been severely pasteurized, e.g. at 80 °C for 30 s. This acts in three ways by

i. Lowering the P^H value.
ii. Increasing the calcium ion concentration.
iii. Raising the colloidal calcium–phosphate content.

Changes in the Biochemical Characteristics During Cheese Ripening

Cheese ripening is characterized by three main biochemical processes, namely

A. Glycolysis
B. Lipolysis
C. Proteolysis

The products of these primary reactions undergo numerous modifications and interactions. An overview of the principal biochemical changes is:

Glycolysis

Most (about 98%) of the lactose in cheese milk is removed through whey as lactose or lactic acid. However, fresh cheese curd contains 1–2% lactose, which is normally metabolized to L-lactic acid by the *Lactococcus* starter within a day for most varieties or a few weeks for cheddar. The L-lactate is racemized to DL-lactate by NSLAB within about three months in the majority of the cheese varieties, and very small amount is oxidized to acetic acid. The rate of oxidation is dependent on the oxygen content of cheese, and hence, on the permeability of the packaging material.

Cheese varieties are made by using *Streptococcus salivarius* ssp. *Thermophilus* and *Lactobacillus* spp. as starters, e.g. Swiss-type and Mozzarella. The metabolism of lactose is more complex than in cheese, in which a *Lactococcus* starter is used. In these varieties, the cheese curd is cooked to 52–55 °C. The cooking temperature is above the growth temperature for both starter components. As the curd cools, the *Streptococcus*, being the more heat-tolerant of the two starters, begins to grow, utilizes the glucose moiety of lactose, and produces L-lactic acid. Galactose is not metabolized and accumulates in the curd. When the curd has cooled sufficiently, the *Lactobacillus* spp. grow, and if a galactose-positive strain is used, it metabolizes galactose, producing DL-lactate. Where a *Lactobacillus* of galactose negative strain is used, galactose accumulates in the curd and can participate in maillard browning, especially during heating, which is undesirable.

Swiss-type cheeses are ripened at about 22 °C for a period to encourage the growth of *Propionibacterium* spp., which use lactic acid as an energy source, producing propionic acid, acetic acid, and CO_2.

$$CH_3CHOHCOOH \rightarrow CH_3CH_2COOH + CH_3COOH + CO_2 + H_2O$$

Acetic and propionic acids likely add to the flavor of Swiss cheese while CO_2 produces large characteristic eyes.

In surface mold ripened cheeses, e.g. Camembert and Brie, *Penicillium camemberti* grows on the cheese surface, metabolizes lactic acid as an energy source causing the P^H to increase.

In the surface smear ripened cheese, e.g. Munster, Limburger etc., the cheese surface is colonized first by yeast which catabolize lactic acid causing the P^H to increase, and then by *Brevibacterium linens*, the characteristic microorganism of the surface smear but which does not grow below P^H 5.8, and various other micro-organisms, including *Micrococcus* and *Coryneform* bacteria.

Lipolysis

Some lipolysis occurs in all cheese; the resulting fatty acids add to the cheese flavor. Lipolysis is rather limited in most of the cheese varieties and is caused mainly by the limited lipolytic activity of the starter and nonstarter lactic acid bacteria, perhaps with a contribution from indigenous milk lipase, especially in raw milk derived cheese varieties.

Extensive Lipolysis occurs in two families of cheese in which fatty acids and/or their degradation products are major contributors to flavor, i.e. certain Italian varieties e.g. Romano and provolone and the blue cheeses.

Blue cheeses undergo very extensive Lipolysis during ripening, up to 25% of all fatty acids may be released. The principal lipase in blue cheese is that produced by *Penicillium roqueforti* with minor contribution from indigenous milk lipase and the lipase of starter and nonstarter lactic acid bacteria. The free fatty acids contribute directly to the flavor of blue cheeses but, more importantly, they undergo partial b-oxidation to alkan-2-ones (methyl ketones) through the metabolic activity of molds. The characteristic peppery flavor of blue cheese is due to alkan-2-ones. Under anaerobic conditions, some of the alkan-2-ones may be reduced to the corresponding alkan-2-ols (secondary alcohols), which cause off flavors.

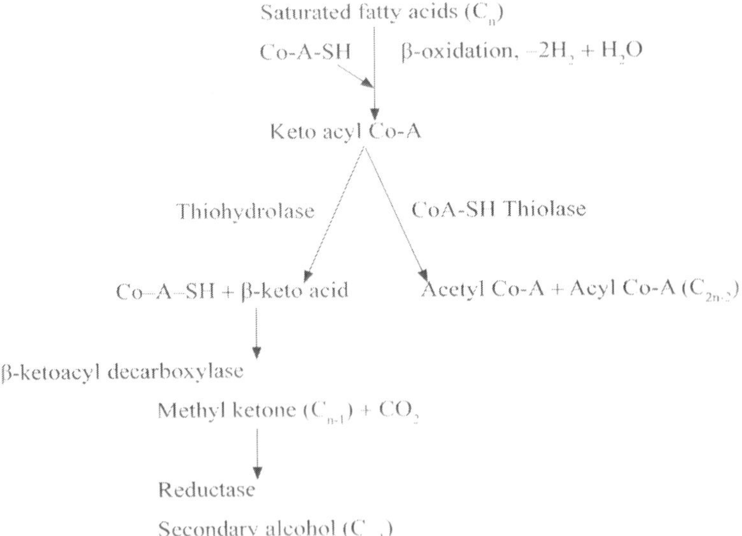

Proteolysis

Among the three main biochemical processes involved in the ripening of most cheese varieties, proteolysis is the most intricate and possibly the most significant. In internal, bacterially ripened cheeses, e.g. Cheddar, Dutch, and Swiss varieties, it is mainly responsible for the textural changes that occur during ripening, i.e. conversion of the tough rubbery texture of fresh curd to the smooth, pliable body of mature cheese. Small peptides and free amino acids enrich the cheese flavor directly and amino acids serve as substrate in several flavor generating reactions, e.g. decarboxylation, deamination and desulfuration. Amino acids may also react chemically with carbonyls via the maillard reaction and strecker degradation, and produce a great diversity of sapid compounds. Excessive amounts of peptides of hydrophobic nature may be produced under certain circumstances and may lead to bitterness which some consumer find very objectionable.

The extent of proteolysis in different cheeses varies from limited (e.g. mozzarella) to moderate (e.g. cheddar and gouda) to very extensive (e.g. blue cheese). The proteolytic products of cheese range from very large polypeptides to amino acids. These are catabolized to a very diverse range of sapid compounds, including amines, acids, and sulfur compounds.

Accelerated Cheese Ripening

Ripening cheese, especially low-moisture varieties, is a slow process. The ripening process is sometimes very unpredictable. Hence, there are economic and technological incentives to accelerate ripening while retaining or improving flavor and textural characteristics.

The main methods for quickening the ripening of cheese are

1. Elevated ripening temperature especially for cheddar which is now usually ripened at 6–8 °C.
2. Exogenous enzymes, usually proteinase and/or peptidase. For several reasons, this approach has had limited success except for enzyme modified cheese.
3. Attenuated lactic acid bacteria, e.g. freeze shocked, heat shocked or lactose negative mutants.
4. Adjunct starter, especially mesophilic lactobacilli.
5. Use of fast lysing starters which die and release their intracellular enzymes rapidly.
6. Genetically modified starters which super produce certain enzymes.

Acid Curd And Rennet Curd

1. Lactic acid produced by the action of a lactic starter culture coagulates milk in an acid curd.

2. Rennet curd is in which milk is coagulated by rennet action in the presence of lactic acid, developed in turn by the starter culture action.
3. Optimum P^{H} for rennet activity is 5.4.
4. Optimum P^{H} of acid curd is 4.6.
5. The coagulum produced by rennet has considerable elasticity and undergoes shrinkage, thereby causing elimination of whey.
6. The coagulum produced by acid curd is not elastic but gelatinous and fragile.
7. The reaction of the rennet curd is close to neutrality; conditions are more favorable for the development of numerous groups of microorganisms.

   In acid curd, the lower P^{H} restricts such growth.

   Rennet curd has large particles/flakes.

   Acid curd has small particles/flakes.

FSSA (2006) Standard Of Various Cheeses

The various cheeses may contain food additives permitted as laid down in FSSA (2006). It shall also conform to the microbiologial requirements as laid down in FSSA (2006). It shall conform to the following requirements:

Product	Moisture	Milk fat on dry basis
Hard pressed cheese	Not more than 39.0%	Not less than 48.0%
Semi hard cheese	Not more than 45.0%	Not less than 40.0%
Semi soft cheese	Not more than 52.0%	Not less than 45.0%
Soft cheese	Not more than 80.0%	Not less than 20.0%
Extra hard cheese	Not more than 36.0%	Not less than 32.0%
Mozzarella Cheese	Not more than 60.0%	Not less than 35.0%
Pizza cheese	Not more than 54.0%	Not less than 35.0%

References

Bandyopadhyay, A. K., Ghatak, P. K., and Ray, P. R. Textbook on Analysis of Milk Products. New Delhi: Kalyani Publishers, 2016.

De, Sukumar. 2004. Outlines of Dairy Technology. New Delhi: Oxford University Press.

Fox, P. E. and McSweeney, P. L. H. Dairy Chemistry and Biochemistry. Madras: Blackie Academic & Professional, 1998.

FSSA. The Food Safety and Standard Act, Professional Book Publishers. New Delhi: FSSA, 2006.

Mathur, M. P., Datta Roy, D., and Dinakar, P. 1999. Text Book of Dairy Chemistry. New Delhi: Indian Council of Agricultural Research. Govt. of India

Varnam A. H. and Sutherland J. P. Milk and Milk Products: Technology, Chemistry and Microbiology. London: Chapman & Hall, 1994.

9

Chemistry of Ice Cream

Commercially, one of the most significant milk products is ice cream. It is a frozen product which is usually made from milk or cream that has been flavored with fruits like peaches or strawberries, and spices like vanilla or cocoa, or with sugar or a substitute. Stabilizers are sometimes combined with food coloring. Ice cream mix typically contains 10.0% milk fat, 11.0% solids-not-fat, 14.0% added sugar, and 0.3% additives like an emulsifier, stabilizer, flavor, and color substances.

Standard of Ice Cream

According to FSSA (2006), ice cream, kulfi, chocolate ice cream or softy ice cream means the product obtained by freezing pasteurized milk prepared from milk and/or other products derived from milk with the addition of sweetening agents and fruit products, egg products, coffee, cocoa, ginger, and nuts. It may also contain chocolate, and bakery products such as cake or cookies as a separate layer and/or coating. It may be frozen hard or frozen to a soft consistency. It shall be free from artificial sweeteners. It shall have a pleasant taste and smell free from flavor and rancidity. It may contain food additives. It shall conform to the microbiological requirements as given in FSSA (2006). The said product shall also conform to the following requirements:

Requirements	Ice cream	Medium-fat ice cream	Low-fat ice cream
Total solids	Not less than 36.0%	Not less than 30.0%	Not less than 26.0%
Wt/vol. (g/l)	Not less than 525	Not less than 475	Not less than 475
Milk fat	Not less than 10.0%	More than 2.5% but less than 10.0%	Not more than 2.5%
Milk protein (Nx6.38)	Not less than 3.5%	Not less than 3.5%	Not less than 3.0%

Dried ice cream mix/dried dessert/confections means that the product is in powder form, which, on addition of the prescribed amount of water, shall give a product conforming to the requirements of the respective products, namely ice cream, medium-fat ice cream, and low-fat ice cream, except for the requirement of weight or volume for both products. The moisture content of the product shall not be more than 4.0%. It may contain food-permitted additives under FSSA (2006).

Frozen dessert/frozen confection means the product obtained by freezing a pasteurized mix prepared with milk fat and/or edible vegetable oils and fats having a melting point of not more than 37.0 °C in combination with milk protein alone or in combination/or vegetable protein products in combination with the addition of sweetening agent and fruit products, egg products, coffee, cocoa, ginger, and nuts. It may also contain chocolate and bakery products such as cake or cookies as a separate layer and/or coating. It may be frozen hard or frozen to a soft consistency. It shall be free from artificial sweeteners. It shall have a pleasant taste and flavor free from off-flavor and rancidity. The product may contain food additives. It shall conform to the chemical and microbiological requirements as given in FSSA (2006). The said product shall also conform to the following requirements:

Requirements	**Frozen dessert/ frozen confection**	**Medium fat frozen dessert/ frozen confection**	**Low fat frozen dessert/frozen confection**
Total solids	Not less than 36.0%	Not less than 30.0%	Not less than 26.0%
Wt/vol. (g/l)	Not less than 525	Not less than 475	Not less than 475
Milk fat	Not less than 10.0%	More than 2.5 % but less than 10.0 %	Not more than 2.5%
Total protein (Nx6.38)	Not less than 3.5%	Not less than 3.5%	Not less than 3.0%

Milk ice or milk lolly means the product obtained by freezing a pasteurized mix prepared from milk and/or other products derived from milk with or without the addition of nutritive sweetening agents, fruit, and fruit products, eggs products, coffee, cocoa, chocolate, condiments, spices, ginger, and nuts; the said product may also contain bakery products such as cake or cookies as a separate layer and/or coating; the said product shall have pleasant taste and smell free from off flavor and rancidity. It may contain food additive as permitted under FSSA (2006). It shall conform to the following requirements:

Total solids (m/m) not less than 20.0%

Milk fat (m/m) not more than 2.0%

Milk protein (Nx6.38) not less than 3.5%

Role of Different Constituents in the Ice Cream Mix

Milk Fat

Milk fat has a significant impact on the quality of ice cream. It contributes to flavor and texture. Not only does the milk fat properly balance the ice cream mix, but also satisfy the legal standard. It enriches and softens the ice cream, giving it a full, rich, creamy flavor. It also controls the whipping rate during processing. Milk fat is a good carrier and synergists for added flavoring compounds and it promotes desirable tactile qualities. The best source of milk fat in ice cream is fresh cream. Other sources are frozen cream, plastic cream, butter, butter oil and condensed milk.

Milk Solids Not Fat (MSNF)

MSNF consists of milk protein, lactose, and minerals, which helps in maintaining the mix consistency and ensures ice cream with a smooth texture. It is essential for the formation and maintenance of small, stable air cells. It also enhances palatability, increases food value, and is economical. They are responsible for parts of freezing point depression and for an increase of the viscosity. Lactose adds a sweet taste to the product while protein contributes to the development of ice cream structure including emulsification, whipping, and water holding capacity. Excessive amounts of MSNF result in salty or cooked flavor and soggy or sandiness.

Sweeteners

Sweeteners make ice cream more palatable to the consumers. Ice cream sweeteners may be cane sugar (sucrose) alone or may be cane sugar plus some corn product. The sugar may be used in dry or liquid form. Sweeteners balance the fitness of the added fat and also improve the texture, palatability, nutritional quality, and flavor. Sweeteners are usually the cheapest source of total solids in ice cream. Sweeteners lower the freezing point and increase the viscosity. The most commonly used ice cream sweeteners are sucrose, dextrose, corn sweetener, inverted sugar, honey, and saccharin.

In the industry, it is now standard procedure to replace all or a portion of sweeteners with glucose sirup or starch hydrolysate sirup. These products result from the hydrolysis of corn starch by either acid, enzyme, or both. The degree of hydrolysis is expressed as dextrose equivalent (DE). DE is a measure of the reducing sugar content of the sirup, calculated as dextrose. These sweeteners contribute a firmer and more chewy body, provide better meltdown, and reduce heat shock potential. These qualities extend the ice cream's shelf life and offer a cheap source of solids. High conversions DE sirups are more economical sources of sweetener than lower-conversion DE sirups because they are sweeter and required less. Higher DE corn sirups decrease freezing point, viscosity, fat agglomeration, and firmness of the product compared to low DE corn sirup. Low-conversion corn sirup is used in ice cream when a thick, chewy texture and resistance to heat shock are desired.

Egg and Egg Products

Egg yolk imparts a characteristic delicate flavor in ice cream. It has a high nutritional value and because of the presence of lecithin it enhances the emulsification qualities of the ice cream mix.

Stabilizer

Ice cream stabilizers are added to the product to stop big ice crystals from forming while it's being stored. Stabilizers function through their abilities to either form gel structures in water or combine the water as water for hydration, thus increasing viscosity. Every stabilizer makes the unfrozen portion more viscous, which restricts molecules migration to crystal nuclei, thereby limiting crystal size.

Stabilizers of vegetable origin should be preferred because no aging is required. They have a negligible effect on food value and ice cream flavor. There are four different kinds of stabilizers in ice cream mix. They are

Protein

Gelatin:Animal origin by nature and a colloidal substance. It can form a gel in the mix but requires 4 h of aging to give the optimum stabilizing effect.

Seaweed Derivatives

Sodium alginate: Alginates are obtained from brown seaweeds, which are found off the coasts of the U.S.A., France, Japan, and North West Scotland and are vegetable in nature. It is obtained as alginic acid or calcium alginate, which is then chemically changed to sodium alginate. No aging is required for sodium alginates, which easily dissolve and develop maximum viscosity as soon as they are in solution. It improves whipping ability and leaves a slightly clear impression in the mouth.

Carrageenan: It is extracted from carrageen (Irish moss), a seaweed growing on the coast of Ireland. It is a sulfated galactose polymer. Carrageenan complexes with milk protein and prevents serum separation or wheezing when the gum stabilizers are used alone.

Cellulose Derivatives

Carboxymethyl cellulose (CMC): CMC is produced from cellulose (wood or cotton). It is highly soluble in water and has high water absorbing power. The grades of CMC are determined by viscosity, and the type of frozen confection being produced influences the choice of stabilizer.

Gums

Guar gum: Guar gum is produced from Guar seeds which is obtained from a plant grown in India.

Locust bean gum: It is mostly available in Cyprus. It is produced from the beans of the locust plant Locust bean gum gives the ice cream a good body and slow melts down. It takes several minutes at relatively high temperature to dissolve, hydrate, and form the gel structure in the mix. It has high water absorption power.

Xanthan gum: It is produced in culture broth media by the microorganism *Xanthomonas campestris* as an exopolysaccharide. This stabilizer has limited use.

Emulsifier

Emulsifiers are added to the ice cream mixture to enhance its whipping capabilities. Improved whipping quality produces ice cream of smooth and dry texture. Emulsifier also facilitate the manufacturing process if ice cream.

Emulsifiers act as surface-active agents and reduce the energy required to maintain the integrity of fat globules. Since they are amphiphilic in nature and possess both hydrophilic and lipophilic properties, they reside at the interface between liquid fat and water. The lipophilic part of the molecules is attracted by the fat, and the remainder of the molecule is hydrophilic and is attracted by the water in the mix. Thus, they reduce the interfacial tension, or force, that exists between the two phases of emulsion and maintain the globule in a stable condition, ensuring that fat agglomeration does not take place. Emulsifier also produces smaller ice crystals along with smaller air cells, which are more evenly distributed and result in smoother ice cream.

Mono- and diglycerides and the sorbitan ester such as polysorbate 80 are used in the manufacture of ice cream.

Monoglycerides are usually derived from hydrogenated vegetable and animal fat where palmitic and stearic acid esters are predominant. They improve the fat dispersion and whipping ability and have a moderate effect on stiffness and melting rate.

Polysorbate emulsifiers are a sorbitan ester consisting of a glucose alcohol (sorbitol) molecule bound to a fatty acid which is mainly oleic acid, with oxyethylene groups added for further water solubility. These emulsifiers are more hydrophilic than the monoglyceride and are effective in imparting dryness, stiffness to ice cream.

Physico-Chemical Effects of Different Mix Processing Operations

Effect of Pasteurization

The main goal of pasteurization is to make sure the product is safe from microbes. But during ice cream mix pasteurization, the heating also affects the physico-chemical properties of the mix. The emulsifier is melted, and heat-activated stabilizers are brought into the colloidal solution. The whey proteins present in the SNF are partially denatured and uncoil, exposing the lipophilic portion of the molecule. The whey proteins begin to act as an emulsifier because of this. The water-binding capacity is increased simultaneously. Denaturation of whey protein also increases the number of binding sites for protein-hydrocolloid interactions and thus enhances the action of stabilizers. Pasteurization improves the quality of ice cream, but excessive heat treatment causes organoleptic deterioration.

Effect of Aging

Aging induces several physico-chemical changes in the ice cream mix. Skim milk powder and stabilizers become fully hydrated, and the emulsifier-induced desorption process of proteins continues. During aging, substantial crystallization of fat also takes place. High-melting-point triacylglycerols crystallize first and are located most closely to the fat globule surface. The crystallization process further involves low-melting-point triacylglycerols, creating multiple-shelled fat globules with a core of liquid fat. Poor-quality ice cream is made from unsaturated fat with a relatively low degree of fat crystallization.

Effects of Freezing

One of the crucial steps in making ice cream is freezing the mixture, as it controls the quality, palatability, and yield of the finished product. The final structure of ice cream is determined during the freezing and aeration of the mix. Ice cream mix with the proper amount of coloring and flavoring material is quickly frozen while being agitated to incorporate air in such a way that it produces and controls the formation of small ice crystals, which is necessary to give smoothness to the finished ice cream. Usually, freezing is done in a scraped-surface heat exchanger. Mix is pumped in to the freezer together with a metered amount of air. As it passes through the freezer, the temperature is lowered as scraper blades sweep the mix over the refrigerated well.

When ice cream is partially frozen to a certain consistency, it is drawn from the freezer into packages and quickly transferred to cold storage where freezing and hardening process is complete without agitation. Fast freezing is essential for a smooth product because ice crystals that are frozen quickly are smaller than that freeze slowly. Therefore, it is desirable to freeze and draw from the freezer in a short time as possible.

Effect of Hardening

Freezing changes the mix to a flowable semi-frozen mass with half the water into ice which is then ready for hardening at about −25 °C to form a solid mass. Rapid hardening is necessary to keep the ice crystal size small. If the temperature fluctuations are allowed to occur in the storage room, during transportation, or in dealer cabinets, small ice crystals will melt down when the temperature rises, and the water will be deposited as ice on the large crystals or re-freezed. In this manner of repeated temperature fluctuations, known as heat shock, a coarse texture results from the development of large individual or agglomerated ice crystals.

References

Bandyopadhyay, A. K., Ghatak, P. K., and Ray, P.R. Text Book on Analysis of Milk Products. New Delhi: Kalyani Publishers, 2016.

De, Sukumar. Outlines of Dairy Technology. New Delhi: Oxford University Press, 2004.

FSSA. The Food safety and Standard Act, Professional Book Publishers. New Delhi: FSSA, 2006.

Mathur, M. P., Datta Roy, D., and Dinakar, P. Text Book of Dairy Chemistry. New Delhi: Indian Council of Agricultural Research. Govt. of India, 1999.

Varnam A. H. and Sutherland J. P. Milk and Milk Products: Technology, Chemistry and Microbiology. London: Chapman & Hall, 1994.

10

Chemistry of Concentrated Milk

Concentrated milk products are obtained through partial removal of water which increases the dry matter content of milk. The increase in dry matter content in concentrated milk is proportional to the degree of concentration. The milk concentrate may be prepared from either full-cream whole milk or fat-free skim milk.

Depending on the nature of raw material and other non-milk additives concentrated milks are classified into four broad categories.

1. Sweetened condensed milk
2. Evaporated milk or unsweetened condensed milk
3. Sterile milk concentrate
4. Reconstituted concentrated milk

Sweetened Condensed Milk

One of the earliest dairy products to be produced industrially is sweetened condensed milk. According to FSSA (2006), sweetened condensed milk is the product obtained by partial removal of water from the milk of cow/buffalo with the addition of sugar or a combination of sucrose with other sugars or by any other process that leads to a product of the same composition and characteristics. The fat and/or protein content of the milk may be adjusted by the addition and/or withdrawal of milk constituents in such a way as not to alter the whey protein to casein ratio of the milk being adjusted. It shall have a pleasant taste and flavor and be free from off-flavor and rancidity. It shall be free from any substance foreign to milk. It may contain food additives. It shall conform to the microbiological requirement as laid down by FSSA. The said product shall also conform to the following requirements (Table 1):

Table 1: FSSA (2006) of sweetened condensed milk.

Product	Milk fat	Milk solids	Milk protein in milk solids not fat
Sweetened condensed milk	Not less than 9.0 % m/m	Not less than 31.0% m/m	Not less than 34.0% m/m
Sweetened condensed skimmed milk	Not more than 1.0% m/m	Not less than 26.0% m/m	Not less than 34.0% m/m
Sweetened condensed partly skimmed milk	Not less than 3.0% m/m	Not less than 28.0% m/m	Not less than 34.0% m/m
Sweetened condensed high-fat milk	Not less than 16.0% m/m	Not less than 30.0% m/m	Not less than 34.0% m/m

In the tropical region, sweetened condensed milk is used in cooking, tea, and coffee. Additionally, the production of chocolate and candies uses it.

Sweetened condensed milk is highly viscous because of its high sugar content. The product is a little glassy in appearance because of fat globules, casein micelles, and dissolved substances. Lactose contributes to the product's turbidity. A large part of lactose crystallization occurs due to supersaturation. Sweetened condensed milk's high sugar content raises the osmotic pressure to a point where the majority of the microorganisms are destroyed. Heat treatment of the product is not done after packaging as its high sugar content preserves it for a long shelf life.

Evaporated Milk

Evaporated milk is sterilized, concentrated, homogenized milk. It is light in color. The product can be stored without refrigeration, has a long shelf life, and is safe for the user.

According to FSSA (2006), Evaporated milk means the product obtained by partial removal of water from the milk of cow and/or buffalo by heat or any other process that leads to a product of the same composition and characteristics. The fat and protein content of the milk may be adjusted by addition and/or withdrawal of the milk constituents in such a way as not to alter the whey protein to casein ratio of the milk being adjusted. It shall have a pleasant taste and flavor free from off-flavor and rancidity. It shall be free from any substance foreign to milk. It may contain food additives permitted by FSSA (2006). It shall conform to the microbiological requirements as laid down by FSSA. The said product shall also conform to the following requirements (Table 2).

Table 2: FSSA (2006) of evaporated milk.

Product	Milk fat	Milk solids	Milk protein in milk solids not fat
Evaporated milk	Not less than 8.0% m/m	Not less than 26.0% m/m	Not less than 34.0% m/m
Evaporated partly skimmed milk	Not less than 1.0% but not more than 8.0% m/m	Not less than 20.0% m/m	Not less than 34.0% m/m
Evaporated skimmed milk	Not more than 1.0% m/m	Not less than 20.0% m/m	Not less than 34.0% m/m
Evaporated high-fat milk	Not less than 15.0% m/m	Not less than 27.0% m/m	Not less than 34.0% m/m

Nutritional Status of Concentrated Milk

Heat treatment during processing causes loss of nutrients during processing of concentrated milk but there is a little change in the biological value (BV) of proteins. The concentrated milk's amino acid composition is nearly identical to that of the raw milk used to make it. Because of maillard browning, concentrated milk may lose up to 20% of its lysine content, which is found in raw milk. Vitamin losses after evaporation are almost similar to the loss that takes place after the pasteurization of raw milk. Loss of vitamins continues during storage of canned concentrated milk. Storage temperature is the prime controlling factor in the loss of vitamins in concentrated milk. In concentrated milk, sterilization causes a 20% reduction in vitamin C content; this loss rises with longer storage times and higher temperatures. Vitamin B_1 and B_2 may fall by approximately 30% during one year of storage of concentrated milk.

Effect of Concentration on Milk Properties

The concentration of milk during the processing of different concentrated milk products has several effects on the physico-chemical properties of milk. The physico-chemical changes are more pronounced in sweetened condensed milk as compared to evaporated milk because of the lower concentration factor. The major physico-chemical changes that take place during the processing of concentrated kinds of milk are

i. Water activity (a_w) of the final product is lowered.
ii. Viscosity increases. The final product becomes non-Newtonian due to the high-concentration factor.

iii. Density and refractive index increase but thermal conductivity decreases.

iv. Hygroscopicity of the product increases.

v. There is a decrease in the diffusion coefficient

vi. Colligative properties of milk such as osmotic pressure, freezing point depression, and boiling point elevation increases.

vii. Electrical conductivity increases.

viii. The increase in ionization and solubility of salts increases because of an increase in ionic strength with a corresponding decrease in the activity coefficient of ionic species.

ix. Thermodynamic activity of the solutes increases.

x. Dissolved substances become supersaturated and may precipitate.

xi. Protein conformation may change. Tendency toward compact structure formation and association increases.

xii. Casein micelle size increases. This phenomenon is contributed by coalescence rather than swelling and the increase is less if the milk is preheated sufficiently to form casein-whey protein complex.

Other Changes During Evaporation

Significant alterations occur in fat globules during the processing of evaporated milk. Fat globules break apart into several tiny globules with distinct surface layers. The tiny fat globules are therefore very vulnerable to lipolysis unless lipase is deactivated. Moreover, evaporation eliminates dissolved gases from milk and lessens off flavors by eliminating volatile substances.

Heat Stability of Concentrated Milk

Concentrated milk's heat stability is crucial for business. The term heat stability of evaporated milk refers to the relative resistance of the milk to coagulations in the sterilizer. Webb Bell and Deysher defined heat stability as "the time necessary to initiate coagulation at 115 °C (239°F)." In the manufacturing process of concentrated products, heat stability is crucial. It is required to adjust heat stability before sterilizing when making recombined evaporated milk from anhydrous milk fat and skimmed milk powder. This is a result of how easily milk can now be transported and stored in the forms of skim milk powder and anhydrous milk fat.

Sommer and Hart (1926) found that apart from the destabilizing effect of albumin, heat stability is regulated by the ratio of calcium and magnesium to phosphate and citrate. Adjustment of the ratio by the addition of main phosphates permits successful product sterilization in most cases. The main rationale behind the commercial sterilization of evaporated milk is still the "salt balance" theory.

Heating concentrated milk eventually causes coagulation, and stability is typically measured in terms of heat coagulation time (HCT), which is the amount of time required for milk to coagulate visibly at a specific temperature. Coagulation of evaporated milk during sterilization has been regarded as a problem during processing. The salt balance theory helped in the development of the procedures to control the coagulation industrially. Subsequently, the reaction between β-lactoglobulin and the κ-casein of the casein micelle was thought to be the most significant factor in changing susceptibility to coagulation. It also involved the stability of the colloidal caseinate-phosphate complex and the changes brought about by heating.

Concentrated (evaporated) milk hardly can be sterilized without preheating the milk before concentrating. The heat stability maximum is shifted to lower pH values and becomes higher. The initial raw milk pH before preheating also affects the heat stability of the evaporated milk, its optimum being, for instance, 6.45. The positive impact of preheating must be primarily attributed to the actions of β-lactoglobulin; the inclusion of this protein in milk reduces its heat stability; however, this reduction can be largely countered by appropriate preheating.

The short duration of ultra-high temperature preheating results in a "concentrate" that is very stable to further heat processing, but its low viscosity makes it unsuitable for producing satisfactory evaporated milk. The stability of recombined, evaporated milk is considerably improved if the original milk is preheated at 120 °C for 2 min. The acid side of the natural pH range is where maximum stability occurs. It may be preferable to use a combination of NaH_2 PO_4 and Na_2 HPO_4 instead of HCI when adjusting the pH.

Age Thickening and Gelation of Concentrated Milk

All concentrated kinds of milk have issues with age thickening and gelation. Particular problems are associated occasionally with canned, sterilized, and evaporated milk. The issue of age thickening is not caused by the disintegration of proteins, but rather by the creation of a three-dimensional network that is made up of thread-like structures connecting aggregated casein micelles. Cold storage of the concentrate before sterilization strongly promotes age thickening and gelation and this phenomenon is attributed to a physico-chemical transformation of the casein micelles. Two distinct stages are observed in age thickening and the rate of age thickening is lower at sterilization heat loads. The two distinct stages of age thickening are

1. Increase in viscosity accompanied by "Spotty lump" formation where the fluidity of milk is intact after stirring.
2. Irreversible gelation accompanied by syneresis and flocculation after stirring.

Age thickening and, eventually, gelation are thought to be the primary changes that occur in sweetened condensed milk during storage. Sweetened condensed milk is far more concentrated than evaporated milk. However, it does not thicken markedly faster with age. It is generally considered that adding more sucrose prevents age-thickening. Sucrose increases the Ca^{2+} activity. One distinction with evaporated milk is that viscosity does not initially decrease before age thickening. The viscosity in sweetened condensed milk increases almost linearly with time. The following main factors affect age thickening in sweetened condensed milk:

1. The variation in the milk type and season occurs among batches of milk.
2. Higher preheating treatment yields higher initial viscosity, and gel can form earlier. Hence, UHT heating is now generally applied.
3. Sugar addition at later in evaporating process results in lesser age thickening.
4. Higher concentration factor gives more age thickening.
5. The impact of added salts varies greatly and is contingent upon the addition stage. Salts are added up to 0.2%. Adding a little quantity of sodium tetrapolyphosphate (e.g. 0.03%) mostly delays age thickening considerably, whereas adding more may have the opposite effect.
6. Age thickening considerably increases with storage temperature.

Lactose Crystallization in Sweetened Condensed Milk

Sweetened condensed milk contains around 38 to 45 g of lactose per 100 g of water. The solubility of lactose at room temperature is about 20 g per 100 g of water, however, because sweetened condensed milk contains sucrose, the solubility of lactose is reduced by approximately half. Due to the high viscosity, nucleation will be slow and only a few nuclei will be formed per unit volume of milk, leading to large crystals.

In the absence of special precautions, the product will produce a sizable number of large crystals. These crystals settle and are responsible for a sandy mouth feel. The crystals may not be large enough to be felt individually in the mouth, but they may still be large enough to leave an uneven impression. To avoid this, they should be smaller than about 8 μ m in length. Preventing crystallization is not possible and, accordingly, a large number of crystals should be obtained.

Satisfactory results can be reached by using seed lactose. Adding 0.03% seed lactose represents 0.004 times the amount of lactose to be crystallized. The final size of the crystals in the product should not exceed 8 μm. Consequently, the lactose would contain enough seed crystals (one per crystal to be formed) if its crystal size does not exceed about $(0.004 \times 8^3)^{1/3}$, i.e. 1.25 μm. Such tiny crystals can be made by intensive grinding of α-lactose hydrate.

References

Bandyopadhyay, A.K., Ghatak, P.K., and Ray, P.R. *Text Book on Analysis of Milk Products*. New Delhi: Kalyani Publishers, 2016.

De, Sukumar. *Outlines of Dairy Technology*. New Delhi: Oxford University Press, 2004.

Fox, P.E. and McSweeney, P.L.H. *Dairy Chemistry and Biochemistry*. Madras: Blackie Academic & Professional, 1998.

FSSA. *The Food Safety and Standard Act*. New Delhi: Professional Book Publishers, 2006.

Mathur, M. P., Datta Roy, D., and Dinakar, P. *Text Book of Dairy Chemistry*. New Delhi: Indian Council of Agricultural Research Govt. of India, 1999.

Varnam A. H. and Sutherland J.P. *Milk and Milk Products: Technology, Chemistry and Microbiology*. London: Chapman & Hall, 1994.

11

Chemistry of Dried Milk

The process of making dried milk involves eliminating water from milk, primarily through spray or roller drying. It is regarded as a significant sector of the dairy industry and is made from whole cow, buffalo, or standardized milk. Skim milk powder (SMP) is made from milk after the fat has been removed whereas whole milk powder (WMP) is made from whole cow or buffalo milk (Table 1).

Table 1: Composition of dried milk (% w/w).

Types of milk	Moisture	Fat	Protein	Lactose	Ash
WMP	2.0	27.5	26.4	38.2	5.9
SMP	3.0	0.8	35.9	52.3	8.0

Source: Webb, B. H., Johnson, A., and Alford 1978. *Fundamental of Dairy Chemistry.*

FSSA (2006) Standard Of Milk Powder

According to FSSA (2006), Milk powder is the product obtained by partial removal of water from the milk of cows and/or buffalo. The fat and/or protein content of the milk may be adjusted by the addition and/or withdrawal of milk constituents in such a way as not to alter the whey protein to casein ratio of the milk. It shall be of uniform color and shall have pleasant taste and flavor, free from off-flavor and rancidity. It shall also be free from vegetable oil/fat, mineral oil, thickening agents, added flavor, and sweetening agents. It may contain food additives. It shall conform to the microbiological requirements prescribed by FSSA (2006). The said product shall also conform to the following requirements:

Requirement	Whole milk powder	Partly skimmed milk powder	Skimmed milk powder
Moisture	Not more than 4.0% m/m	Not more than 5.0% m/m	Not more than 5.0% m/m
Milk Fat	Not less than 26.0% m/m	Not less than 1.5% m/m and not more than 26.0% m/m	Not more than 1.5% m/m
Milk protein in milk solids not fat	Not less than 34.0% m/m	Not less than 34.0% m/m	Not less than 34.0% m/m
Titrable acidity (mL 0.1N NaOH/10 g solids not fat)	Not more than 18.0	Not more than 18.0	Not more than 18.0
Insolubility index	Not more than 2 mL	Not more than 2 mL	Not more than 2 mL
Total ash on dry weight basis	Not more than 7.3%	Not more than 8.2%	Not more than 8.2%

Physico-Chemical Changes During Manufacture of Dried Milk

Physico-chemical changes that are observed during the processing of dried milk are

a. Maillard reaction, an amino-sugar reaction,
b. heat-induced coagulation, and
c. lipid oxidation

Water activity, high storage temperatures, longer than usual storage times, high product moisture content, high processing temperatures, and container headspace are some of the variables that affect the Maillard reaction in dried milk.

Casein and whey proteins are the major proteins present in milk. Unlike casein, whey proteins are heat-labile. Whey protein fractions present in milk are α-lactalbumin, β-lactoglobulin, bovine serum albumin, immunoglobulin, euglobulin, and lactoferrin. β-lactoglobulin and κ-casein interact during heat treatment. When milk is refrigerated for longer than 24 h, it forms γ-casein and protease-peptone fractions.

Dried milk is susceptible to lipid oxidation which is largely affected by the degree of unsaturation, fatty acid composition, and storage temperature, exposure to light, moisture content, and head space of the container. The flavor of dried milk has largely been affected by lipid oxidation.

Chemical Changes in Dried Milk During Storage

Microorganisms cannot flourish in dried milk due to low water activity and storage defects are purely chemical. Two major types of deterioration are well recognized, Maillard browning and fat decomposition. The rate of these changes is influenced by composition, quality of milk, heat treatment, moisture content, metallic contamination, packaging method, and condition of storage (temperature, humidity, and light).

Maillard Browning

Maillard browning is an amino-sugar reaction. Dried milk, during long storage, gradually develops a brown color which is caused by Maillard-type interaction between the free amino groups of milk proteins and the aldehyde group of lactose, resulting in loss of available lysine, nutritive value, and off-flavor development.

The chemistry of Maillard browning is very complex and the reactions that take place are as follows:

1. Initial stage (Products are colorless without absorbance of UV radiation at 280 nm).

   A. Sugar-amino condensation.

   B. Amadori rearrangement.

2. Intermediate stage (products are colorless or yellow, with strong absorption of UV light).

   A. Sugar dehydration.

   B. Sugar fragmentation.

   C. Amino acid degradation.

3. Final stage: (products are highly colored)

   A. Aldol condensation

   B. Aldehyde-amino polymerization

Glucosamine is first formed through a reaction between lactose and amino groups, primarily lysine residue. An amadori rearrangement occurs with the formation of N- substituted 1-amino-1-deoxy-2-ketose. The Amadori compound is dehydrated in the second stage to produce reductones and furfural, fission compounds, strecker degradation compounds. The final stage consists of the conversion of the furfural, fission compound reductones into melanoidin pigments with further involvement of amines. Pyrazines, a strongly flavorful compound, can be produced by the reaction of amino acids with the breakdown products of Strecker degradation, namely dicarbonyls.

Fat Decomposition

This may be divided into oxidative, hydrolytic rancidity, and ketonic rancidity

Oxidative Rancidity

The primary storage flaw in whole milk powder and, to a lesser extent, in the residual fat content of skim milk powder is fat oxidation, which results in the production of a flat and finely marked oxidized flavor. The oxidized flavor becomes noticeable by sensory tests in roughly three months when the dry milk/ whole milk powder is held at 21–24 °C under air-packed conditions.

Factors of fat oxidation

The following factors affect the oxidation in dried milk

1. An increase in storage temperature causes a rise in the rate of fat oxidation.
2. An increase in acidity causes a significant increase in the rate of fat oxidation.
3. Milk powder's fat oxidation is induced by light. Transparent packaging should not be used to avoid this.
4. Iron and copper function as a catalyst to speed up the fat oxidation process induction phase.

Method of prevention

1. Inert gas packaging: the oxygen in the container is eliminated by nitrogen gas packaging. This process can prolong the storage life and virtually eliminate oxidation.
2. Addition of antioxidants: Antioxidants significantly reduce the oxidation of fat. According to FSSA (2006), whole milk powder may contain 0.01% more butylated hydroxy anisole (BHA) by weight of the final product.
3. High preheating temperature: elevated milk temperature during preheating could potentially regulate fat oxidation and enhance the shelf life of milk powder. It is assumed to be associated with the formation of traces of protein decomposition products containing sulfhydryl compounds which act as anti-oxidant and prolong the induction period.

Hydrolytic Rancidity

The hydrolysis of fat produces free fatty acids like butyric acid, which is what causes hydrolytic rancidity. Lipase enzyme is responsible for this deterioration. Milk itself contains lipase or it may be produced by microorganisms. Since lipase can be destroyed by adequate pasteurization and the customary high preheat treatment, the defect is extremely rare.

Ketonic Rancidity

Dried milk may develop a stale off-flavor, most characteristically "coconut," a toffee flavor during storage. These kinds of off-flavors are more common in whole milk powder than in skim milk powder. The off-flavor is caused by a rearrangement of 9-decanoic acid in milk fat to form delta-deca lactone in small quantities. Lowering the storage temperature delayed the development of stale flavor. The appearance of stale flavor is often accelerated by overheating the milk and exposing it to heat after drying.

References

Bandyopadhyay, A. K., Ghatak, P. K., and Ray, P. R. Text Book on Analysis of Milk Products. New Delhi: Kalyani Publishers, 2016.

De, Sukumar. Outlines of Dairy Technology. New Delhi: Oxford University Press, 2004.

Fox, P. E. and McSweeney, P. L. H. Dairy Chemistry and Biochemistry. Madras: Blackie Academic & Professional, 1998.

FSSA. The Food Safety and Standard Act. New Delhi: Professional Book Publishers, 2006.

Mathur, M. P., Datta Roy, D., and Dinakar, P. Text Book of Dairy Chemistry. New Delhi: Indian Council of Agricultural Research Govt. of India, 1999.

Varnam A. H. and Sutherland J. P. Milk and Milk Products: Technology, Chemistry and Microbiology. London: Chapman & Hall, 1994.

12

Chemistry of Ghee

One of the most consumed and well-known milk products in the Indian subcontinent is ghee. It is a by-product of tradition, prepared both at organized dairy plants and the household level for a well-established market. Ghee is traditionally made from the milk of cows or buffalo. The tetrapyrrole pigments biliverdin and bilirubin are responsible for the distinctive color of buffalo ghee. On the other hand, ghee is golden yellow due to the presence of carotenoids in cow's milk. Ghee's composition changes according to the kind and feeding habits of the animal and can differ from place to place and season to season.

In addition to, being a substantial source of fat-soluble vitamins, essential fatty acids, and other growth-promoting components, ghee is a rich source of energy. Its value as food for humans is enormous. Additional attributes related to medicine are present in ghee. Ghee can improve one's physical and mental attributes and treat eye conditions like ulcers. Ghee is used in religious and ceremonial rites in addition to being consumed.

Composition

Ghee is a complex lipid composed of mixed glyceride which makes up 99–99.5% of its total content. In addition, it has trace amounts of calcium, magnesium, phosphorus, iron, copper, and fat-soluble vitamins, tocopherols, carbonyls, hydrocarbons, carotenoids, free fatty acids, phospholipid, sterols, and sterol esters (Table 1).

Table 1: Chemical composition of ghee

Characteristics	**Requirements**	
	Cow	**Buffalo**
Milk fat	99 to 99.5%	
Moisture	Not more than 0.5%	
Unsaponifiable matter		
a. Carotene (µg/g)	3.2–7.4	-
b. Vitamin A (I.U/g)	19–34	17–38
c. Tocopherol (µg/g)	26–48	18–37
Free fatty acid (%oleic acid)	max 2.8 (Agmark)	
Charred casein, salts of copper	Traces	
and iron, etc.		

Source: Outlines of Dairy Technology (2004).

Role of Different Chemical Constituents

The majority of the sterols in ghee are cholesterol. The structure or pattern of arrangement of fatty acids in the triglyceride molecule is an important aspect of compositional studies of ghee. The chemical characteristics of ghee are significantly influenced by the glyceride structure. The glyceride composition of cow ghee is distinctly different from that of buffalo ghee. The proportion of high melting point triglyceride is much higher in buffalo ghee (average 8.7%) than in cow ghee (average 4.9%). This is because buffalo ghee contains a higher percentage of long-chain fatty acids, primarily stearic and palmitic acids. For this reason, buffalo ghee is a little bit harder than cow ghee.

Cow ghee contains a higher amount of long-chain triglycerides as compared to buffalo ghee (Table 2). Because buffalo ghee has a higher butyric acid content than cow ghee, it has a higher proportion of short-chain triglycerides (Table 3).

Table 2: Major and minor constituents of cow and buffalo ghee.

Characteristics	Cow	Buffalo
Saponifiable constituents		
Triglycerides[a]		
Short chain (%)	45.3	37.6
Long chain (%)	54.7	62.4
Trisaturated (%)	40.7	39.0
High melting (%)	8.7	4.9
Partial glycerides		
Diglycerides (%)	4.5	4.3
Monoglycerides (%)	0.6	0.7
Phospholipids (mg%)	42.5	38.0
Unsaponifiable constituents		
Total cholesterol (mg%)	275.0	330.0
Lanosterol (mg%)	8.27	9.32
Lutein (μ/g)	3.1	4.2
Squalene (μ/g)	62.4	59.2
Carotene (μ/g)	0.0	7.2
Vitamin A (μ/g)	9.5	9.2
Vitamin E (μ/g)	26.4	30.5
Ubiquinone (μ/g)	6.5	5.0
Flavor components		
Total carbonyls (μM/g)	8.64	7.2
Volatile carbonyls (μM/g)	0.25	0.035

[a]Based on the percentage of total glycerides (Sharma 1981).

Table 3: FSSA (2006) standard of ghee.

S.No	Name of the state/ union territory	Butyro refractometer reading at 40 °C	Minimum Reichert value	Percentage of FFA as oleic acid (max)	Percentage of moisture (max)
1.	Andhra Pradesh	40.0–43.0	24	3.0	0.5
2.	Andaman and Nicobar Islands	41.0–44.0	24	3.0	0.
3.	Arunachal Pradesh	40.0–43.0	26	3.0	0.5
4.	Assam	40.0–43.0	26	3.0	0.5
5.	Bihar	40.0–43.0	28	3.0	0.5
6.	Chandigarh	40.0–43.0	28	3.0	0.5
7.	Dadra and Nagar Haveli	40.0–43.0	24	3.0	0.5
8.	Delhi	40.0–43.0	28	3.0	0.5
9.	a. Goa b. Daman and Diu	40.0–43.0 40.0–43.5	26 24	3.0 3.0	0.5 0.5

10.	**Gujarat**				
	a. Areas other than cotton tract areas	40.0–43.5	24	3.0	0.5
	b. Cotton tract areas	41.5–45.0	21	3.0	0.5
11.	**Haryana**				
	a. Areas other than cotton tract areas	40.0–43.0	28	3.0	0.5
	b. Cotton tract areas	40.0–43.0	26	3.0	0.5
12.	Himachal Pradesh	40.0–43.0	26	3.0	0.5
13.	Jammu and Kashmir	40.0–43.0	26	3.0	0.5
14.	Jharkhand	40.0–43.0	28	3.0	0.5
15.	**Karnataka**				
	a. Areas other than the Belgaum district	40.0–43.0	26	3.0	05.
	b. Belgaum district	40.0–44.0	26	3.0	0.5
16.	Kerala	40.0–43.0	26	3.0	0.5
17.	Lakshadweep	40.0–43.0	26	3.0	0.5
18.	**Madhya Pradesh**				
	a. Areas other than cotton tract areas	40.0–44.0	26	3.0	0.5
	b. Cotton tract areas	41.5–45.0	21	3.0	0.5
19.	**Maharashtra**				
	a. Areas other than cotton tract areas	40.0–43.0	26	3.0	0.5
	b. Cotton tract areas	41.5–45.0	21	3.0	0.5
20.	Manipur	40.0–43.0	26	3.0	0.5
21.	Meghalaya	40.0–43.0	26	3.0	0.5
22.	Mizoram	40.0–43.0	26	3.0	0.5
23.	Nagaland	40.0–43.0	26	3.0	0.5
24.	Orissa	40.0–43.0	26	3.0	0.5
25.	Pondicherry	40.0–43.0	26	3.0	0.5
26	Punjab	40.0–43.0	28	3.0	0.5
27.	**Rajasthan**				
	a. Areas other than Jodhpur District	40.0–43.0	26	3.0	0.5
	b. Jodhpur District	41.5–45.0	21	3.0	0.5
28.	Tamil Nadu	41.0–44.0	24	3.0	0.5
29.	Tripura	40.0–43.0	26	3.0	0.5

30.	Uttar Pradesh	40.0–43.0	26	3.0	0.5
31.	Uttaranchal	40.0–43.0	26	3.0	0.5
32.	**West Bengal**				
	a. Areas other than Bishnupur sub division	40.0–43.0	28	3.0	0.5
	b. Bishnupur sub division	41.5–45.0	21	3.0	0.5
33.	Sikkim	40.0–43.0	28	3.0	0.5

Baudouin test shall be negative.

Chemical Constants of Ghee

i. **Saponification value**: It is defined as milligrams of KOH required to saponify 1 g of fat or oil. For fat or oil containing short-chain fatty acids or low molecular weight, the saponification number is high, and vice versa. It gives a clue about the molecular weight and size of the fatty acid in the fat or oil.

ii. **Iodine value**: It is defined as the number of grams of iodine taken up by 100 g of fat or oil. Iodine number is a measure of the degree of unsaturation of the fatty acid. Since the quantity of iodine absorbed by the fat or oil can be measured accurately, it is possible to calculate the relative unsaturation of fats or oil.

iii. **Reichert-Meisel value (RM value)**: This is a measure of the volatile soluble fatty acids. It is defined as the number of milliliters of 0.1 N alkali required to neutralize the soluble volatile fatty acids contained in 5 g of fat. The determination of the Reichert-Meisel number is important because it helps to detect adulteration in butter and ghee. Reichert-Meisel value is reduced when animal fat is used as an adulterant in butter or ghee.

iv. **Polanski value:** It is possible to contaminate ghee by adding nonvolatile, insoluble fatty acids (by adding animal fat). To test this, one can ascertain the Polanski number. It is defined as the number of milliliters of 0.1 N potassium hydroxide solution required to neutralize the insoluble fatty acids (not volatile with steam distillation) obtained from 5 g of fat.

v. **Acetyl number:** It is defined as the amount in milliliters of potassium hydroxide solution required to neutralize the acetic acid obtained by saponification of 1 g of fat or oil after acetylation. Some fatty acids contain hydroxyl groups. To determine the proportion of these, they are acetylated using acetic anhydride. This results in the introduction of acetyl groups in the place of free hydroxyl groups. The acetic acid in combination with fat can be determined by titration of the liberated acetic acid from acetylated fat or oil with standard alkali. Acetyl number is thus a measure of the number of hydroxyl groups present in fat or oil.

vi. **Acid number:** It is defined as the milligram of potassium hydroxide required to neutralize the free fatty acids present in one gram of fat or oil. The quantity of free fatty acids in fat or oil is indicated by the acid number. The age of the fat or oil increases its content of free fatty acids.

References

Bandyopadhyay, A. K., Ghatak, P. K., and Ray, P. R. Text Book on Analysis of Milk Products. New Delhi: Kalyani Publishers, 2016.

De, Sukumar. Outlines of Dairy Technology. New Delhi: Oxford University Press, 2004.

FSSA. The Food Safety and Standard Act. New Delhi: Professional Book Publishers, 2006.

Mathur, M. P., Datta Roy, D., and Dinakar, P. Text Book of Dairy Chemistry. New Delhi: Indian Council of Agricultural Research Govt. of India, 1999.

Rangappa, K. S. and Acharya, K. T. Indian Dairy Products. Bombay: Asia Publishing House, 1974.

Sharma, R.S. 1981.Ghee: A resume of recent research. J. Food Sci. Technol. 18(2)70-76.

13

Chemistry of Traditional Indian Dairy Products

Being an essential component of the rich Indian past, traditional dairy products and a sophisticated classification of confections have played a significant role in the Indian economy. Depending on how it is processed, it has significant social, cultural, and economic significance. These products have been developed over a long period using a combination of experience, traditional wisdom, and culinary skills. India is emerging as the highest milk-producing country in the world with an annual production of 221.1 million tons in the year 2021–22 (Source: Basic Animal Husbandry Statistics, MoFAHD, DAHD, GoI). An estimated 50% of the milk produced in India is used to make a range of traditional dairy products, including paneer, dahi, ghee, khoa, and shrikhand. The product is not only well-established in the Indian market, but it also has significant export potential due to the large Indian diaspora living all over the world.

All native milk products that have developed over time using locally accessible technologies are referred to as traditional Indian dairy products or Indian indigenous milk products. The Indian masses' economic, social, religious, and nutritional well-being are greatly influenced by traditional dairy products.

Classification of Traditional Indian Dairy Products

Traditional dairy products in India are typically made, packaged, and/or sold using time-honored methods without the use of machinery or meticulous documentation of the procedure. These products do, however, have several technological advantages, including their widespread appeal, reduced cost of production, straightforward manufacturing process, adoption at the rural level or collection center, large profit margin, and abundance of employment opportunities. These products are categorized according to the manufacturing principle. The broad classification of traditional Indian dairy products is given in Table 1.

Table 1: Classification of traditional Indian dairy products.

Manufacturing type	Base material	End products
Acid coagulated	Chhana Paneer	Rasgulla, sandesh, pantoa, cham-cham, chhana murki, chhana-podo, Sitabhog culinary dishes
Heat desiccated	Khoa	Peda, burfi, kalakand, milk cake, gulab jamun, rabri
Fermented/cultured	Chakka	Dahi, mishti doi, lassi, shrikhand, shrikhand wadi
Fat rich		Ghee
Frozen	-	Kulfi/malai
Cereal-based	-	Kheer, payasam, Phirni

Chhana

Chhana is a heat-acid coagulated milk product made by draining the whey after boiling the whole milk to a high temperature. It has a creamy-white look, a compact, spongy body, a sweet-acidic flavor, and is used as the foundation for several well-known Indian sweets. The body and texture characteristics are mainly influenced by the temperature and amount of coagulation. An approximate of 4% of the nation's milk production is thought to be turned into chhana.

According to FSSA (2006), chhana is a product obtained from cow or buffalo milk or a combination thereof by precipitation with sour milk, lactic acid, or citric acid. Milk solids may be used in the preparation of this product. It shall not contain more than 70% moisture, and the milk fat content shall not be less than 50% of the dry matter. Chhana, when sold as low-fat chhana, shall not contain more than 70% moisture, and the milk fat content shall not exceed 15% dry matter.

Chemistry of Chhana

Chhana is made from milk that has been coagulated by acid and heat, which alters the casein's chemical and physical properties. From the normal colloidal dispersion of discrete casein micelles, casein entrains fat and coagulates serum protein with whey to form large structural aggregates, or curd. The curd then separates from the milk serum as a flocculent. A part of calcium and phosphate are lost from the casein micelles during boiling and subsequent coagulation.

Many factors, including milk type, fat content, homogenization, calcium level, additives, type and quantity of coagulant, temperature of coagulation, milk pH, and stirring speed, affect the quality and chemical composition of chhana. Since cow milk produces small grains, a smooth texture, and a soft body that are ideal for making sweets, it is preferred when making chhana. Buffalo milk chhana has a greasy, coarse texture and a slightly hard body. The hard body and coarse texture of buffalo milk chhana are caused by a higher casein concentration, primarily in the form of larger-sized micelles in the micellar form, a higher proportion of high-melting triglycerides, and a higher calcium content. Chhana contains 51–53% moisture, 24–29% fat, 14–17% protein, 2.1– 2.3% lactose, and 2.0–2.1% ash, depending on whether it is prepared from cow, buffalo, or mixed milk.

Factors Affecting Physical Quality of Chhana

The physical quality of chhana is influenced by various factors such as milk's mineral content, fat content, colostrum presence, milk adulteration, coagulant type, strength, temperature, and processing method.

The amount of fat in the milk affects the chhana's final quality. Chhana must have a minimum fat content of 4% in cow's milk and 5% in buffalo's milk for satisfactory quality. A softer chhana with more fat loss in the whey would result from a higher fat level.

The amount of calcium in milk directly affects how hard chhana is. While sodium chloride, sodium acetate, and sodium citrate did not affect the body or texture of chhana, the addition of calcium or magnesium chloride results in hard chhana.

Colostrums tend to give the coagulation mass in milk a pasty texture and a deeper yellow color. Chhana made with colostrum milk is unsuitable for use in confections.

When milk is contaminated with starch, it coagulates into a gelatinous mass that is unfit for use in confections.

During chhana making, organic acids such as citric acid, lactic acid, or their salts, lemon juice, or sour whey are typically used as coagulants. Lactic acid (0.5–0.75%) produces a granular texture suitable for rasgulla making, while citric acid (1–2%) produces a doughy product suitable for sandesh making. Calcium lactate (4%) yields a smooth, soft-bodied, and flavorful chhana that is ideal for making sandesh. When rasgulla is prepared, a coagulant with a low acid strength typically produces a soft body and smooth texture; when a coagulant with a high acid strength is used, the body becomes relatively hard and the texture becomes less smooth.

Coagulation temperature exerts a significant influence on physico-chemical attributes. A temperature range of 80–85°C and pH 5.4 for cow milk and 70–80°C and pH 5.7 for buffalo milk will give better quality and yield. Raising the coagulation temperature causes the chhana to become harder, more granular, and grainy. It also results in less moisture content but a higher percentage of milk solids, which include fat, protein, lactose, and ash. The lower temperature imparts a sticky chhana with a slow drainage problem.

While rapid stirring increases the moisture content and hardens the chhana, slow stirring is generally preferred to prevent foam formation and lessen coagulation mass shattering. Once more, delayed straining yielded a higher yield of chhana while maintaining a relatively smooth and soft texture. A softer body, smoother, stickier, and less chewy texture are the results of delayed whey drainage after coagulation, and this is more tolerable in terms of overall quality.

Trypsin coagulation during the proteolysis of buffalo milk produced a chhana with a very smooth and soft texture, minute-sized micelles, and a body that was ideal for making rasgulla. When 0.3% sodium citrate is added to buffalo milk, a chhana that resembles cow milk chhana is produced. Suitable chhana is made from a blend of 75% buffalo milk and 25% cow milk, making for a good rasgulla.

The chemical reactions that transpire after manufacture and during storage of chhana significantly modify its physical and chemical characteristics, making it unsuitable for consumption.

The product stored in vegetable parchment paper had a shelf life of two, three, and twelve days at roughly 37, 24, and 70°C, respectively. The storage temperature guides the character of spoilage. Low temperatures cause a stale flavor to develop along with a heavy growth of mold on the surface. Conversely, when the product is stored at high temperatures, it takes on an unpleasant taste and sour smell, and the surface of the chhana starts to grow mold.

Sandesh

Sandesh, a popular chhana-based dessert in eastern India, is typically made by heating a mixture of ground chhana and cane sugar in a shallow pan while constantly stirring and scraping the mixture until the desired consistency and flavor develop. After cooling, the mixture is molded into the desired shape. Sandesh is broadly classified into three main categories, namely, soft Sandesh (naram pak), hard sandesh (karapak), and kachagolla.

Table 2: Chemical composition of sandesh (% w/w).

Constituents	**Soft grade**	**Hard grade**	**Kachagolla**
Moisture	23.4	11.6	33.89
Fat	23.1	21.8	15.50
Protein	26.1	20.5	12.75

Sucrose	47.9	53.0	35.75
Ash	1.71	1.69	1.45
Acidity (% lactic acid)	0.88	0.82	0.70

Source: Sen and Rajorhia 1989.

Since it produces a final product with a soft body, smooth texture, and small grains, cow milk chhana is typically chosen when making sandesh. Due to compositional variations, buffalo milk chhana has a hard body, a coarse texture, and larger grain sizes.

Rasgulla

One of West Bengal's most well-known chhana-based sweet is rasgulla. Snow-white in hue, rasgulla has a smooth texture and a spongy, chewy body. Cow milk chhana works best for this preparation because buffalo milk chhana lacks the necessary body and texture and is hard and spongy. The process of making rasgulla involves coagulating hot, boiled milk with acid and then draining the whey. The coagulum is then cooked in sugar syrup after kneading it with some additives and making small balls. Finally, these cooked balls are soaked in sugar syrup again.

Table 3: Chemical composition (% w/w) of rasgulla.

Constituents	Range	Average
Total solid	52.98–58.35	55.98
Fat	4.65–8.67	6.53
Protein	7.23–8.53	7.97
Sucrose	33.98–37.67	35.45
Ash	0.61–1.37	0.93
Additives	2.11–6.51	5.10

Other popular chhana-based traditional Indian sweets are Rasmalai, Cham-cham, pantooa, Chhana-murki, khirmohan, lalmohan, sitabhog, mihidana, chhapdo, churpi, and jalbhara. These dairy-based sweets are produced in a very disorganized, regionally specific manner that necessitates intensive research and development for process improvement and large-scale marketing.

Paneer

India's well-known heat-acid coagulated milk product, paneer, is made by coagulating standardized milk at a predetermined temperature with permitted acids. The resulting coagulum is pressed and filtered to produce a mass of curd that can be sliced. Paneer has a smooth texture and a firm, close, cohesive, and spongy body. It is mostly made with buffalo milk and used in a wide range of recipes. Also known as Indian cottage cheese, paneer is typically sold in blocks or slices. It is estimated that paneer is made from 5% of the milk produced in India, and its production is increasing at a rate of 13% per year.

According to FSSA (2006), paneer is the product obtained from cow or buffalo milk or a combination thereof by precipitation with sour milk, lactic acid, or citric acid. Milk solids may be used in the preparation of this product. It shall not contain more than 70% moisture, and the milk fat content shall not be less than 50% of the dry matter. Paneer, when sold as low-fat chhana, shall not contain more than 70% moisture, and the milk fat content shall not exceed 15% of dry matter.

Table 4: Chemical composition (%w/w) of paneer.

Constituents	**Cow**	**Buffalo**
Moisture	52–54	50–52
Fat	24–26	28–30
Protein	16–19	13–15
Lactose	2.0–2.2	2.2–2.4
Ash	2.0–2.3	1.9–2.1

Chemistry of Coagulation Process in Paneer

The normal colloidal dispersion of discrete casein micelles, in which whey and coagulated serum proteins are entrapped together with milk fat, gives rise to large structural aggregates of casein during the coagulation process. The primary alterations that occur during coagulation are

1. The gradual removal of tri calcium phosphate from casein's surface and its transformation into soluble calcium salt and mono calcium phosphate.
2. Calcium is gradually removed from calcium hydrogen caseinate to produce free casein and soluble calcium salt.
3. The colloidal particles in the milk system become isoelectric or have no net electric charge when the pH falls. In such a situation, the casein precipitates and forms a coagulum, rendering the dispersion unstable.

The development of the typical rheological features of paneer may be the result of a high concentration of heat-induced protein-protein interaction. It is thought that when heated together, β-lactoglobulin and κ-casein interact through sulfhydryl disulfide inter changes. It starts at 65°C, rises to 83% at 85°C, and falls to 76% at 99°C.

Factors Affecting Paneer Quality

The paneer quality is influenced by many factors which are discussed below:

i. **Quality of milk:** It is recommended to homogenize cow's milk in order to increase yield and raise the organoleptic score of paneer. Products made from acidic milk with a titratable acidity of 0.2–0.23% LA are of lower quality.

ii. **Type of milk:** It is considered that buffalo milk is better for making paneer than cow milk. Paneer made from buffalo milk has superior body, texture, and spongy qualities.

iii. **Heat treatment of milk:** The process of heating milk to 80°C for 5 min and then cooling it to 70°C before coagulation has an impact on the sensory and microbiological quality of paneer.

iv. **Type of coagulant:** Different coagulants, such as citric acid solution, 1-day-old whey, 48-h-old sour whey, and whey cultured with *Lactobacillus acidophilus*, are added to milk to prepare paneer. The production of paneer has also involved the use of additional coagulants, including phosphoric acid, tartaric acid, and acetic acid.

v. **Strength of coagulant:** For buffalo milk paneer, 1% citric acid yields the highest quality. Good-quality paneer was also produced using 2% citric acid and cow's milk.

vi. **Coagulation temperature:** It has been suggested that the optimal coagulation temperature is 70°C. But high-quality paneer can also be made with both cow and buffalo milk at a coagulation temperature of 80°C.

Khoa

Khoa is a heat-desiccated milk product. In India, khoa is made from about 7–8% of all milk produced. Traditionally, it is made by continuously stirring milk that has been desiccated in an open, shallow pan until a coagulation mass forms, which is then worked up to form a pat. It serves as the foundation for a wide variety of milk-based sweets, including gulabjamun, peda, burfi, and milk cakes. Three distinct types of khoa, namely, pindi, dhap, and danadar are available in the country. These varieties differ in quality, texture, composition and specific usage. Depending on whether it is made from cow, buffalo, or mixed milk, khoa typically has 20–25% moisture, 25–27% fat, 17–20% protein, 22–25% lactose, 3.6–3.8% ash, and 100–130 ppm of iron.

According to FSSA (2006), khoa by whatever variety of names it is sold such as pindi, danedar, dhap, mawa, or kova means the product obtained from cow or buffalo or goat or sheep milk (or milk solid) or a combination thereof by rapid drying. The milk fat content shall not be less than 30% on dry matter basis of the finished product. It may contain citric acid not more than 0.1% by weight. It shall be free from added starch, added sugar and added coloring matter.

Buffalo milk is typically chosen over cow milk during khoa making because it produces a larger yield and has a more palatable flavor, body, and texture.

Chemistry of Khoa

- Khoa is basically a matrix of heat-denatured milk protein that contains moisture, fat, lactose, and other trace elements. During desiccation, reactions with other milk constituents and between proteins take place.
- Lactose and protein that undergo Maillard browning and caramelization are significant in imparting typical flavor and desirable sensory characteristics to the khoa
- The main reaction in khoa preparation is the heat denaturation and coagulation of milk protein, particularly the serum proteins and their interaction with casein and other constituents. Early in the boiling process, the protective qualities of the other colloids are destroyed, and the majority of albumin and globulins are nearly denatured.
- Casein is also irreversibly denatured from a colloidal state to a non-dispersible state. The denaturation is accelerated by the incorporation of air and frothing during stirring. Total heat coagulation of protein occurs when the boiling mixture thickens to a buttery consistency in the pan.
- Fat globules in milk, undergo subdivision and disrupt the fat globule membrane as a result of heat, vigorous agitation, and scraping during khoa manufacture. Almost half of the globular fat is released as free fat, the extents of which depend upon the type and amount of fat content in the milk.
- Formation of free fat during the transition from the liquid phase to the semi-solid phase is very significant for the development of typical flavor, body, and texture of khoa.
- The short-chain fatty acids decrease, the medium-chain fatty acids increase, and the long-chain fatty acids are unaffected during khoa preparation.
- Lactose is present in khoa mostly as supersaturated solution. As lactose is highly viscous and causes protein coagulations, it typically does not crystallize out. However, large crystals of lactose are found in ordinary khoa.

- A portion of milk salt is precipitated during preparation of khoa. Iron content, probably derived from heating pan, is raised considerably.
- Vitamin A, thiamine, riboflavin, ascorbic acid, nicotinic acid and folic acid are decreasing in the course of heating of milk for making khoa.
- The deterioration of khoa is attributed to unhygienic methods of processing and subsequent improper handling, followed by lactose breakdown, proteolysis, and lipolysis due to activities of the mixed flora during storage.
- Being aerobic, molds cover the entire product kept in the ambient conditions producing an abnormal color, flavor, and appearance.

A portion of milk salt is precipitated during the preparation of khoa. The iron content, probably derived from the heating pan, is raised considerably. Vitamin A, thiamine, riboflavin, ascorbic acid, nicotinic acid, and folic acid are decreasing in the course of heating milk for making khoa.

Different khoa-based traditional Indian dairy products are Kalakand, peda, burfi, gulabjamun, and rabri.

Traditional Indian Fermented Dairy Products

Dahi

One of the most well-known traditional fermented dairy products in India is dahi, or curd. In almost every region, it is a part of the daily food and is also used as a base for the preparation of chakka, shrikhand, and lassi. A good-quality dahi has a pleasant flavor, a sweet aroma, and a firm, smooth consistency. The cut surface is solid, free of cracks and gas bubbles, and the surface is glossy and smooth.

According to FSSA (2006), dahi or curd is the product obtained from pasteurized or boiled milk by souring, natural or otherwise, by a harmless lactic acid or other bacterial culture. Dahi may contain additional cane sugar. Milk solids may also be used in preparation of this product. It should have the same percentage of fat and SNF as the milk from which it is prepared. When dahi or curd other than skimmed milk dahi, is sold or offered for sale without any indication of the class of milk, the standard prescribed for dahi prepared from buffalo milk shall apply.

The dahi may be classified into various types depending upon the type of milk used such as whole milk dahi, skim milk dahi. It is also classified on the basis of acidity and addition of sugar such as: sweet dahi (acidity $< 0.7\%$), sour dahi (acidity $> 0.7\%$), and sweetened dahi. The type of milk used and the conditions of manufacturing affect the composition of dahi. The average composition of dahi is given in the Table 5.

Table 5: Composition (%w/w) of dahi.

Constituents	Dahi (whole milk)	Dahi (skimmed milk)
Water	85–88	50–52
Fat	5–8	28–30
Protein	3.2–3.4	13–15
Lactose	4.6–5.2	2.2–2.4
Ash	0.70–0.75	1.9–2.1
Lactic acid	0.5–0.11	

Source: Outlines of Dairy Technology (2004).

Biochemical and Physico-Chemical Changes in Dahi

The following biochemical changes take place during processing of dahi from milk:

1. Break-down of lactose:
   - Lactose is hydrolyzed to glucose and galactose by starter organism in the first step of fermentation during the preparation of dahi.
   - The liberated glucose enters the classical Embden Meyer Hoff pathway (EMP). Galactose produced by hydrolysis cannot enter the EMP pathway directly but follows a longer route and ultimately enter the EMP pathway via glucose–1–phosphate.
   - Several enzymes act on each of the simpler before arriving at the lactic acid stage with the production of several acid end product.
   - The amount of lactic acid produced may range from 75 to 95% of the total acidity, the rest being volatile acid.
   - The lactic acid causes an acid flavor first, followed by the combination with calcium in casein to form calcium lactate, setting free the casein and coagulating it when the isoelectric point (4.6) is reached..
   - The fermentation is accompanied by gelling of the solids, principally the proteins and syneresis as a form of thin exudates of clear whey on the surface of dahi.
2. Changes in nitrogen content:
   - The total nitrogen content of milk remains more or less unchanged during fermentation.
   - Non-protein nitrogen, albumin nitrogen, ammonia, and dialyzable nitrogen changes appreciably during fermentation.

- The increase in the non- protein nitrogen during fermentation is mainly caused by the breakdown of protein.

3. Changes in salt content:
   - The mineral content of milk changes considerably during the processing of dahi.
   - Soluble calcium and phosphorus are increased during souring. Citric acid completely disappears during fermentation.
   - All the calcium is present in the ionic form at pH 4.6, and all the phosphorus, except the casein-conjugated one, exists as ionic phosphate.
4. Changes in vitamin content:
   - The type of organisms used in the fermentation process determines the vitamin content of dahi.
   - Fermenting milk with *streptococcus lactis* and *streptococcus cremoris* results in increase in thiamine, riboflavin and folic acid and decrease in vitamin A.
   - Riboflavin, folic acid, and cyanocobalamin (vitamin $B_{12)}$ content in dahi increases with the incorporation of *Propionibacterium shermanii* in lactic culture
   - *Streptococcus cremoris*, *Streptococcus diacetylactis*, *Streptococcus thermophilus* results in increase in riboflavin and folic acid and decrease in nicotinic acid in dahi when used as starter culture.
5. Changes in physico-chemical properties: The electrical charge on the fat particles is neutralized during fermentation, causing the globule to coalesce and rise to the top. The iso-electric point (pH 4.5–4.6) of the fat globule in milk is slightly higher than that of the washed globule (pH 4.3) from cream. A pure culture is able to neutralize the charge on the fat globule at a lower level (1%) of acidity than a mixed starter (1.3–1.5%).

Mishti Doi

A well-liked fermented dairy product from eastern India, particularly in West Bengal, is mishti doi. It has a smooth texture, firm consistency, a light brown to cream color, and a pleasant aroma. For mishti doi, there isn't an FSSA standard available. A typical Mishti-Doi composition is provided in Table 6. It is made by adding 6–6.5% sugar to milk, either while it is boiling or while it is setting. Prolonged heating of sweetened milk at low temperatures leads to milk solid concentration and the development of a brown color. After cooling, dahi culture is added, and the mixture is then filled into earthen cups. The mixture is then left undisturbed overnight to ferment. Once a firm body forms, low-temperature storage at roughly 40°C is typically followed.

Table 6: Composition of different grades of mishti doi (%w/w).

Constituents	Low fat	Medium fat	High fat
Milk fat	2–3	4–5	8–9
Milk SNF	13–14	11–13	10–11
Sugar	17–19	17–18	17–18
Total solid	32–35	32–36	35–38

Source: Bandyopdhyay *et al.* 2016.

Chakka and Shrikhand

Chakka

Chakka is an intermediate product during the preparation of shrikhand. It is mostly popular in the western region of India. Draining the whey from lactic acid-fermented curd is the first step in making chakka. It is a marble white to pale yellow semi-solid product with good texture, uniform consistency, and a diacetyl aroma. It shall conform to the following requirements as per FSSA (2006):

Table 7: Requirements of chakka (as per FSSA 2006).

Requirements	Chakka	Skimmed milk cream chakka	Full chakka
Total solids, percent by weight	Min 30	Min 20	Min 28
Milk fat (on dry basis) percent by weight	Min 33	Min 5	Min 38
Milk protein (on dry basis) percent by weight	Min 30	Min 60	Min 30
Titratable acidity (as lactic acid) percent by weight	Max 2.5	Max 2.5	Max 2.5
Total ash (on dry basis) percent by weight	Max 3.5	Max 5.0	Max 3.5

Chakka when sold without any indication shall conform to the standard of chakka.

Shrikhand

Served as a dessert or snack, shrikhand is a fermented and sweetened milk product that is particularly well-liked in western India. Shrikhand is a semi soft, sweetish sour product made from concentrated dahi. The consistency is influenced to great extent by the moisture, fat and sugar levels. The following requirements are mandatory for shrikhand in accordance with the FSSA (2006).

Table 8: Requirements of shrikhand (as per FSSA 2006)

Requirements	Shrikhand	Full cream shrikhand
Total solids percent by weight	Not less than 58.0	Not less than 58.0
Milk fat (on dry basis) percent by weight	Not less than 8.5	Not less than 10.0
Milk protein (on dry basis) percent by weight	Not less than 9.0	Not less than 7.0
Titratable acidity (as lactic acid) percent by weigh	Not more than 1.4	Not more than 1.4
Sugar (sucrose) (on dry basis) percent by weight	Not more than 72.5	Not more than 72.5
Total ash (on dry basis) percent by weight	Not more than 0.9	Not more than 0.9

For fruit shrikhand, protein, not less than 6.0.

In traditional method, buffalo milk or mixed cow milk is boiled and after cooling to room temperature (30–35°C). After that, lactic culture (dahi) is added, and it is left to incubate for six to eight hours. When the curd is firmly set (acidity 0–1.0%), it is placed in a muslin cloth and hang on a bag for drainage of whey for 6–8 h. The curd is intermittently squeezed to facilitate whey drainage. The resulting solid mass, known as chakka, is combined with sugar, thoroughly kneaded, and then rubbed through muslin cloth to produce a smooth, sour and sweet product. Permitted colors, flavors and fruits, nut, spices are also added to provide variety (Table 9).

Table 9: Chemical composition (%w/w) of shrikhand.

Parameters	Percentage
Moisture	40–45
Fat	5–6
Protein	7–8
Lactose	8–9
Sucrose	40–42
Ash	0.45–0.55

Source: Bandyopdhyay et al. 2016.

Lassi

Lassi is a popular refreshing beverage in the northern and western states of India. Lassi, or buttermilk, is produced as a by-product during the preparation of butter from dahi by the indigenous method. Dahi is churned, with frequent additions of water, until the butter granules are formed. The diluted, beaten curd remaining after the butter is called lassi or butter milk. It is desirable to homogenize the product for improved body and texture.

However, the ingredients of lassi differ from location to location and are influenced by the fat level of dahi, the amount of dilution during churning, and the churning efficiency. The gross chemical composition of lassi is given in Table 10.

Table 10: Chemical composition (%w/w) of lassi.

Parameters	Percentage
Fat	1–2
Protein	2–3
Sucrose	11–12
Titratable acidity (% Lactic acid)	0.4–0.5

References

Bandyopadhyay, A. K., Ghatak, P. K, and Ray, P. R. Text Book on Analysis of Milk Products. New Delhi: Kalyani Publishers, 2016.

De, Sukumar. Outlines of Dairy Technology. New Delhi: Oxford University Press, 2004.

FSSA. The Food Safety and Standard Act. New Delhi: Professional Book Publishers, 2006.

Mathur, M. P., Datta Roy, D., and Dinakar, P. Text Book of Dairy Chemistry. New Delhi: Indian Council of Agricultural Research. Govt. of India, 1999.

Rangappa, K. S. and Acharya, K. T. Indian Dairy Products. Bombay: Asia Publishing House, 1974.

14

Chemistry of Fermented Milk

The terms "cultured dairy foods," "cultured dairy products," and "cultured milk products" are other terms for fermented dairy products. Lactobacillus, Lactococcus, Leuconostoc, and other lactic acid bacteria, along with yeasts in certain products, are used to ferment milk to make these dairy products. The product's shelf life is extended through fermentation, which also improves the product's taste and milk's digestibility. There is a wide variety of fermented milk products available, each with unique flavors and attributes. The following categories apply to fermented milk products, based on the type of fermentation.

Different Types of Fermented Milk Products

Mesophilic lactic fermentation	Cultured buttermilk
	Cultured cream
	Film jolk
	Scandinavian ropy milks
Thermophilic lactic fermentation	Yogurt
	Acid buttermilk
Therapeutic lactic fermentation	Acidophilus milk
	Yakult products
	Acidophilus–Bifidus (AB) yogurts
Lactic/yeast fermentation	Kefir
	Koumiss

Changes in Nutrients Resulting from Fermentation

Fermentation lowers the lactose content in fermented milk products; however, if the starting milk is concentrated or fortified, the finished product's lactose content may be higher than that of liquid milk.

The composition and content of protein and total amino acids are comparable to those of raw milk. Casein degradation is small compared to cheese, accounting for not more than 1% of the total protein. While some products of casein degradation are used by the starter microorganisms, others build up as free amino acids in the fermented kinds of milk. The amounts of free amino acids that build up differ depending on the species. Generally, prolinc is increased in large quantities, followed by serine, alanine, valine, leucine, and histidine. As a result of continuous starter culture activity, the level of free amino acids increases during storage. A limited degree of lipolysis occurs during the manufacture of fermented milk. The amount varies depending on the type of species, but the pattern and concentration of free fatty acids are both impacted. During fermentation and the breakdown of amino acids, trace amounts of volatile fatty acids are also produced.

The activity of the starter culture during fermentation has a significant impact on the vitamin content, which is also influenced by the thermal processing and starting milk composition. Vitamins are initially metabolized and then synthesized by starter microorganisms. The levels of thiamin, riboflavin, nicotinic acid, pantothenic acid, or biotin are somewhat impacted; however, the concentrations of biotin and pantothenic acid are typically slightly decreased, while those of the other vitamins may be increased. During fermentation, folic acid content is increased 100% and choline content typically increases as well. Vitamin B_{12} is required by the *Lb. delbrueckii* ssp. *bulgaricus* in yogurt and levels are reduced.

Impact of Fermented Milk on Nutrient Bioavailability and Digestibility

Compared to milk and other milk products, fermented milk is easier to digest and the specific rate of protein digestion in yogurt is twice that of raw milk. Protein is partially broken down in fermented milk which makes it more digestible. Fermented milk may improve the body's ability to absorb calcium as well as other minerals and trace elements. Because fermented milk has a shorter intestinal transit time than raw milk, it absorbs iron less readily, but phosphorus is more bioavailable.

Flavor of Fermented Milk

Starter culture metabolites are largely responsible for the distinctive flavor of fermented kinds of milk. Numerous compounds are generated that could potentially enhance flavor. The substances are:

- **Non-volatile acids:** Lactic, pyruvic, oxalic, succinic
- **Volatile acids:** Formic, acetic, propionic, butyric
- **Carbonyl Compounds:** Acetaldehyde, acetone, acetoin or diacetyl
- **Miscellaneous:** Amino acids and/or constituents formed by thermal degradation of protein, fat, and lactose.

All fermented milk contains lactic acid, which is a key flavor component. The amount of lactic acid present can determine how acceptable it is, with excess lactic acid producing flavor defects.

Depending on the starter culture employed, acetaldehyde and diacetyl have different relative importance.

Yogurt

Yogurt and its derivatives have become one of the most popular and widely consumed acidified foods in recent years. A good digestibility, high dietetic value, and mild acidic taste as well as stable quality have contributed to this popularity

Yogurt is milk fermented at 37–40 °C with the intervention of specific lactic cultures. Normally yogurt culture is thermophilic and the most widely used varieties are *Streptococcus salivarius sub sp. thermophilus* and *Lactobacillus delbruccki sub sp. bulgaricus.* The acidity of the product varies from 0.7 to 1.1% lactic acid with approximate p^H 4.0–4.2. Variations of the traditional type of yogurt are found in fluid yogurt, frozen yogurt, fruit yogurt, flavored yogurt, dried yogurt, low lactose yogurt, low-calorie yogurt, and "Bioghurt" or "Biogarde."

Fruit yogurt has crushed fruit, jam, and sugar added while artificial flavoring and sugar or sweetening agents to prepare flavored yogurt. Fluid yogurt can be considered broken-clot yogurt with low viscosity. Frozen yogurt has the appearance of ice cream but is similar to yogurt in chemical composition Dried yogurt is obtained in the Middle East by drying under the sun. Low-lactose yogurt is obtained by hydrolysis of lactose with β-D-galactosidase. Low-calorie yogurt is yogurt in which the energy contained is reduced from 250 to 300 kJ/100 g to 170 kJ/100 g. Combination of typical acidifying bacteria with, e.g. *Bifidobacterium bifidum* has resulted in products such as "Bioghurt" and "Biogarde".

Biochemical Changes in Yogurt

All species of the genus *Pediococcus*, many of *Streptococcus*, and *Lactobacillus* are capable of carrying out homolactic fermentation in which glucose is metabolized to pyruvic acid and then to lactic acid via the glycolytic or Embden-Meyerhof-Parnas pathway (EMP). The amount of lactic acid formed is 1.8 mol/mol of glucose, accompanied by small quantities of acetic acid, ethyl alcohol, and carbon dioxide. The galactose can be converted to glucose-6-phosphate by the intervention of uridine diphosphogalactose-4-epimerase passing through-1-phosphate. This was proved for the first time by Leloir.

Recent reviews indicate that selected starter cultures augment several changes that enhance the nutritional value of milk products. Yogurt proteins have been reported to possess a higher *in vitro* digestibility and higher protein quality as estimated by the biological value and gross protein value. Yogurt has a lot of bacterial cells in it, so some of the protein comes from the cells themselves. During yogurt production, amino acids are also released so that the amount that exists free in the yogurt than the amount in the original milk.

The bacteria can modify casein solubility in yogurt which helps to increase the nutritive quality of fermented milk. The sizes of curd that form after fermentation are notably smaller. Smaller curd particles may be better absorbed by the stomach because they have a larger surface area that the digestive juices can attack.

Liposomal enzymes found in the starter culture are nearly solely responsible for any potential lipolytic activity in yogurt. The lipolytic activities that can arise during industrial yogurt production are always weak.

During the storage of yogurt, limited hydrolysis of fat occurs regularly and has been proved by various researchers. Lactic acid bacteria have a hydrolytic action on triglycerides with long chains. The bacteria also have a secondary lipolytic activity that can assume a certain importance only after other microorganisms have reduced the milk fat into more simple compounds.

Beneficial Effects of Yogurt

1. A build-up of lactic fermentation metabolites, such as H_2O_2, antibiotics, and lactocin, can stop the growth of unwanted bacteria.
2. It helps with digestion, promotes healthy intestinal functions, and, when combined with antibiotics, inhibits harmful microorganisms.
3. Different lactic acid bacteria can grow and adapt in the intestine, influencing the growth of other bacterial groups and establishing an equilibrium.
4. There is evidence from several medical publications connecting hypercholesterolemia to coronary heart disease. Products made from fermented milk may reduce plasma cholesterol levels.
5. The gastrointestinal tract's immune system may be boosted by lactic acid bacteria.
6. Toxic compounds produced by the intestinal flora, endogenous substances, and diet may be assimilated and detoxified by them.
7. Bile acids, which may contribute to the colon's carcinogenic process, can be deconjugated by lactic acid bacteria.

References

Bandyopadhyay, A. K., Ghatak, P. K., and Ray, P. R. Text Book on Analysis of Milk Products. New Delhi: Kalyani Publishers, 2016.

De, Sukumar. Outlines of Dairy Technology. New Delhi: Oxford University Press, 2004.

FSSA. The Food Safety and Standard Act. New Delhi: Professional Book Publishers, 2006.

Mathur, M. P., Datta Roy, D., and Dinakar, P. Text Book of Dairy Chemistry. New Delhi: Indian Council of Agricultural Research Govt. of India, 1999.

Varnam, A. H. and Sutherland J.P. Milk and Milk Products: Technology, Chemistry and Microbiology. London: Chapman & Hall, 1994.

15

Milk Adulteration

Food adulteration is a global issue, and because there are insufficient regulations and oversight, developing nations are more vulnerable to it. But in many nations, food adulteration has gone unnoticed. Contrary to popular assumption, milk adulterants can, regrettably, present major health risks that result in fatal diseases. Modern milk adulteration involves increasingly complex techniques, necessitating state-of-the-art research to identify the adulterants. Because milk contains so many nutrients that are needed by both adults and infants, it is referred to as the ideal food. In terms of protein, fat, carbohydrates, vitamins, and minerals, it is among the best sources. Regretfully, milk adulteration is a very common problem worldwide. Potential causes could be the gap between supply and demand, the perishable nature of milk, the low purchasing power of the consumer, and the absence of appropriate detection tests. Because there is insufficient oversight and inadequate law enforcement, the situation is substantially worse in developing and underdeveloped nations. While quantitative detections of adulterants in milk are varied and complex, qualitative detections can be carried out simply using chemical reactions. The type of quantitative detection methods used depends on the milk's adulterants.

Typical Adulterants in Milk

The addition of vegetable protein, milk from different species, whey addition, and watering are the main adulterants in milk that are known to be economically motivated. There is no serious health risk associated with these adulterations. Some adulterants, though, are too dangerous to ignore. Urea, formalin, detergents, ammonium sulfate, boric acid, caustic soda, benzoic acid, salicylic acid, hydrogen peroxide, sugars, and melamine are a few of the main adulterants in milk that have been shown to have serious negative health effects.

The percentage of fat and solid-not-fat (SNF) milk, the amount of protein, and the freezing point are frequently measured to assess milk quality. To raise these metrics, adulterants are added to milk, deceitfully raising the milk's quality. To increase solid-not-fat (SNF), for instance, cane sugar, starch, urea, sulfate salts, and common salts are added. Since urea is a naturally occurring component of raw milk, the Food Safety and Standards Authority of India (FSSAI) Act of 2006 set a maximum limit of 70 mg/100 mL. Milk is supplemented with commercial urea to raise its non-protein nitrogen content. Melamine is added similarlyto increase the protein content fictitiously. To maintain the density of diluted milk while increasing the lactometer reading, ammonium sulfate is added. As preservatives, hydrogen peroxide, salicylic acid, and benzoic acid lengthen the milk's shelf life. Due to the high cost of milk fat, some producers of dairy products and milk remove it to increase their profits; to make up for this, they add non-milk fat, like vegetable oil. The oil is emulsified and dissolved in water with the addition of detergents, creating a frothy solution that resembles milk.

Some adulterants can have serious long-term health effects. Infants may experience renal failure and even die if they consume melamine at levels higher than what is considered safe.Peroxides and detergents have the potential to aggravate digestive issues, including gastritis and intestine inflammation. Due to the effects of undigested starch in the colon, excessive starch in milk can induce diarrhea; however, starch accumulation in the body can be extremely deadly for people with diabetes. The kidneys are overworked when there is urea in milk because they have to remove more urea from the body. Moreover, bicarbonates and carbonates may interfere with hormone signals that control growth and reproduction.

Qualitative Detection Method

Simple chemical reactions based on color are used for the qualitative detection of adulterants in milk. Tests can be conducted in a laboratory that has access to chemical reagents and can take the required safety measures. These methods' main shortcomings are that they are insufficiently precise and only work for a small range of concentrations. Qualitative detections, however, have the advantage of being straightforward, quick, and simple to execute. A few palatable substances are frequently added to milk as adulterants to enhance its flavor. Table 1 discusses their rapid detection in milk.To extend the shelf life and enhance the visual appeal of milk, certain hazardous chemicals are added. A few of those are extremely dangerous and can cause deadly illnesses. Techniques for quickly and easily identifying dangerous chemicals in milk are presented in Table 2. To enhance the milk's appearance, additional blended chemicals, such as soap, detergents, and coloring agents, are occasionally added. Table 3 discusses the qualitative detection of some of those common adulterants in milk.

Table 1: Qualitative and quick identification of various edible adulterants in milk.

Adulterants	Procedure	Observation
Starch	Take 3 mL sample in a test tube. After boiling it thoroughly, cool it to room temperature. Add 1 drop of 1% iodine solution.	The presence of starch is indicated by the blue color.
Sugar	Take 5 mL milk sample in a test tube. Add 1 mL conc. HCl and 0.1 g resorcinol solution. Place the test tube in a water bath for 5 min.	The presence of added sugar is indicated by the red color
Common salt	Take 5 mL of milk sample into a test tube. Add 1 mL of 0.1 N silver nitrate solution. Mix the content thoroughly and add 0.5 mL of 10% potassium chromate solution.	Brick red color indicates milk is free of added salt, yellow appearance indicates the presence of added salts.
Glucose	Take 1 mL of milk sample in a test tube. Add 1 ml of modified Barfoed's reagent. Heat the mixture for exactly 3 min in a boiling water bath. Rapidly cool under tap water. Add one mL of phosphomolybdic acid reagent to the turbid solution.	The presence of glucose is indicated by the deep blue color that appears immediately.
Buffalo milk	Dilute the milk 1/10. Put a drop of diluted milk on the center of a glass slide. Now place a drop of Hansa test serum (duly preserved) on the drop of milk and mix with a glass rod or clean toothpick.	Within thirty seconds, curdy particles appear in milk that contains buffalo milk.

Table 2: Quick qualitative identification of various hazardous substances in milk.

Adulterant	Procedure	Observation
Hydrogen peroxide	A. Add 5 mL of suspected milk sample in a test tube, an equal volume of raw milk, and five drops of 2% solution of paraphenylenediamine.	The presence of hydrogen peroxide as an adulterant is indicated by the blue color.
	B. Take 1 mL milk sample in a test tube add 1 mL of potassium iodide-starch reagent solution and mix well.	When hydrogen peroxide is present as an adulterant, it appears blue.

Formalin	A. Take a 10 mL milk sample in a test tube. Add 5 mL conc. sulfuric acid with a small amount of ferric chloride without shaking.	The presence of formalin is indicated by the appearance of violet or blue color at the intersection of two liquid layers.
	B. Take about 5 mL of milk in a test tube. Take 1 mL of 10% ferric chloride solution in a 500 mL volumetric flask and make up the volume using concentrated hydrochloric acid. Add 5 mL from this solution to the sample in test tube. Keep the tube in boiling water bath for about 3–4 min.	The presence of formalin is indicated by a brownish-pink appearance.
	C. Take 1 mL of sample milk in a test tube. Take saturated solution of 1, 8-dihydroxynaphthalene-3,6-disulphonic acid in about 72% sulphuric acid to make a chromotropic acid solution. Add 1 mL of chromotropic acid solution to the sample in a test tube.	A brownish-pink appearance denotes the formalin's presence.
Ammonium sulfate	A. Take 2 mL of milk in a test tube and add 0.5 ml NaOH (2%) 0.5 mL sodium hypochlorite (2%) and 0.5 mL phenol (5%) Heat in a boiling water bath for 20 sec	Immediately, a bluish color forms that later deepens to blue. Pure milk shows salmon pink color which progressively alters to bluish after 2 h.
	B. Take 10 mL of milk in a 50 mL stoppered test tube. Add 10 mL of TCA solution. Filter the coagulated milk through Whatman filter paper Grade 42. Take 5 mL of clear filtrate. Add a few drops of barium chloride solution.	The appearance of milky-white precipitates suggests that milk has been added with sulfates, such as magnesium, sodium, zinc, and ammonium sulfate.

Urea	A. Take 5 mL milk sample in a test tube. Add an equal volume of 24% TCA to precipitate the fat and proteins of milk. Take 1 mL filtrate and add 0.5 mL 2% sodium hypochlorite, 0.5 mL 2% sodium hydroxide, and 0.5 mL 5% phenol solution, then mix.	When urea is added, a distinctive blue or bluish-greencolor appears, while pure milk stays colorless.
	B. Take 5 mL milk in a test tube, add 0.2 mL urease (20 mg/ mL) Shake well at room temperature and then add 0.1 mL Bromothymol Blue (BTB) solution (0.5%)	The presence of urea in milk is indicated by the appearance of a blue color after 10–15 min. Normal milk has a slight blue color because it naturally contains urea.
	C. Take 5 mL milk sample in a test tube. Add 5 mL p-Dimethyl Amino Benzaldehyde reagent.	A noticeable yellow hue suggests the addition of urea, while a faint yellow hue suggests the presence of urea naturally occurring in milk.
Benzoic and salicylic acid	Take 5 mL milk sample in a test tube. Upon acidification with sulfuric acid, 0.5% ferric chloride solution is added to it drop by drop. Mix it. Five millilitre of milk is taken in a test tube and acidified with concentrated sulphuric acid. 0.5% ferric chloride solution is added drop by drop and mixed well. Development of buff colour indicates presence of benzoic acid and violet colour indicates salicylic acid.	Benzoic acid is indicated by a buff appearance, while salicylic acid is indicated by a violet appearance.
Borax and boric acid	Take 5 mL milk sample in a test tube. Add 1 mL conc. HCl to it. A turmeric paper is dipped and it is dried in a watch glass at 100 °C.	The presence of borax or boric acid is indicated if the turmeric paper turns red.
Nitrate	Take 10 mL sample milk in a beaker. Add 10 mL mercuric chloride solution to it. After mixing, filter through what man No 42 filter paper. Take 1 mL filtrate in a test tube and add 4 mL of diphenyl amine sulfate or diphenylbenzidine reagent.	The presence of nitrates is indicated by the blue color. A sample of pure milk won't turn any color.

Table 3: Quick qualitative identification of various blended adulterants in milk.

Adulterant	Procedure	Observation
Detergent	A. Take 5 mL in a test tube and add 0.1 mL 0.5% Bromocresol Purple (BCP) solution.	The presence of detergent is indicated by a violet appearance. The color of unadulterated milk is faint violet.
	B. Take 5 mL of milk sample into a 15 mL test tube. Add 1 mL of Methylene blue dye solution and 2 mL chloroform. Vortex the contents for about 15 sec and centrifuge at about 1100 rpm for 3 min.	The lower layer's deeper blue hue denotes the presence of detergent in the milk. A comparatively deeper shade of blue in the top layer denotes the absence of detergent in the milk.
Pulverized soap	Take 10 mL milk sample in a test tube. Add equal quantity of hot water to it, then add 1–2 drops of phenolphthalein indicator.	The presence of soap is indicated by the pink color.
Colouring matter	A. Take 10 mL milk sample in attest tube. Add 10 mL diethyl ether. After shaking, allow it to stand.	The presence of added colour is indicated by the yellow appearance in the ethereal layer.
	B. Make the milk sample alkaline with sodium bicarbonate. Dip a strip of filter paper for 2 h.	The presence of annatto is indicated by the red colour that appears on filter paper. This paper turns pink when treated with stannous chloride.
	C. Add a few drops of hydrochloric acid to the milk sample.	A pink appearance suggests the presence of azo dyes.

Synthetic Milk

Recent cross-breeding initiatives and modifications to milch animal feeding practises have resulted in milk from some animals falling short of legal regulations. This has also led to the adulteration of milk, even with harmful substances, to increase the fat and SNF content to meet the legal standard. India's population is constantly growing, which causes a shortage of milk, particularly in the summer. Simultaneously, the proliferation of dairy farms in our country has led to fierce competition in the buying and selling of milk and milk products.This has also led to the birth of synthetic milk.

In the dairy industry, the term "synthetic milk" refers to a novel product made by mixing liquid soap, urea, vegetable oil, sugar, common salt, washing soda, and water. The use of synthetic milk has proliferated throughout India. This artificially produced milk could endanger the lives of young children. It was challenging to identify this milk blended with natural milk. This threat would cause immeasurable harm to the general public as well as the dairy industry as a whole if it is not checked right away.

Negative Impacts of Synthetic Milk

The consumers' health is affected by the extremely harmful chemicals found in synthetic milk, such as urea, detergents, and alkalis, among others. Some of the harmful effects are given as under.

- Swelling of hands and feet (consumption for long time).
- Negative effects on vision can result in blindness.
- Neurological disorders.
- Cardiovascular diseases.
- Liver and kidney diseases may even cause cancer.
- Urea and ammonium compounds can cause acute toxicity, which can manifest in a range of symptoms including muscle tremors, abdominal pain, polyuria, cyanosis, dyspnea, and hyperthermia in cases of advanced toxicosis (Table 4).

Table 4: Physical and chemical disparities between natural and synthetic milk.

Characteristics	Natural milk	Synthetic milk
Colour	White	White
Taste	Palatable	Extremely bitter
Odor	Characteristic milk odor	Soapy becomes distinct on boiling
Texture	No soapy feeling when rubbed between fingers	It feels soapy when rubbed between the fingers.
pH	6.6–6.8	Alkaline 9.0–10.5
Urea Test	Very light because milk naturally contains urea	Distinctly positive
Sugar Test	Negative	Positive
Neutraliser test	Negative	Positive
Vegetable fat test	Negative	Positive

References

Bodhe, A. M. "Detection of Adulteration in Milk"Indian Dairyman 50, no. 7 (1998):19.

FSSAI. Manual of Methods of Analysis of Foods: Milk and Milk Products.New Delhi: Food Safety and Standard Authority of India, Ministry of Health and Family Welfare. Govt. of India, 2015.

Ghatak, P. K. and Bandyopadhyay, A. K. Practical Dairy Chemistry. New Delhi: Kalyani Publishers, 2007.

Mathur, M. P., Datta Roy, D., and Dinakar, P. Text Book of Dairy Chemistry. New Delhi: Indian Council of Agricultural Research Govt. of India, 1999.

Varnam, A. H. and Sutherland J. P. Milk and Milk Products: Technology, Chemistry and Microbiology. London: Chapman & Hall, 1994.

16

Advanced Techniques of Milk Analysis

Electrophoresis

The mobility of ions in an electric field is the basis for the separation method known as electrophoresis. Ions that are negatively charged move toward a positive electrode, while those that are positively charged move toward a negative electrode. Usually, one electrode is biased either positively or negatively, and the other is at the ground for safety reasons. Ions can be distinguished from one another based on their size, shape, and overall charge, which determine their migration rates.

Poly Acrylamide Gel Electrophoresis (PAGE)

Introduction

Electrophoresis is an electrochemical process in which substances with a net electric charge migrate under the influence of an electric current. This migration occurs in an agar gel or a liquid-filled buffered matrix such as cellulose acetate. Positively charged substances travel toward the cathode (negative electrode), while negatively charged substances go toward the anode (positive electrode). Different substances move at different rates depending on their charge. This movement is called electrophoretic mobility.

Proteins are charged molecules. They have a positive as well as a negative charge. Among proteins, particularly caseins, they have different contents. The charge density also varies depending on the molecular weight. Proteins have a net negative charge at a pH above 4.6, and hence they move towards the anode (positive electrode) when subjected to an electric field under alkaline conditions. Due to differences in the charge of individual protein fractions, their migration rates also differ and are hence separated into distinct bands. These bands are stained with the dye amido-black, destained with 7% acetic acid, and observed for the electrophoretic pattern.

Principle

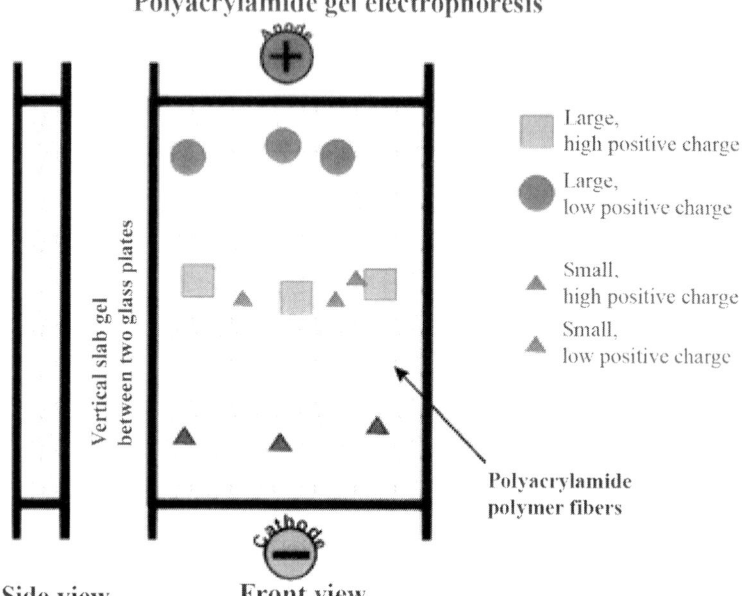

The gel is prepared by polymerizing acryl amide and a small quantity of cross-linking agents $(CH_2{=}CH\ C\ O\ N\ H_2)_2\ CH_2$ methylene bis acryl amide in the presence of a catalyst, ammonium persulfate $(N\ H_4)_2\ S_2\ 0_8$. *Tetramethylethylenediamine*(TEMED) is also present to initiate and control the polymerization. The gel mixture is allowed to polymerize in small tube (or perspex) bottles with rubber cap. A layer of water is placed on top of the gel to ensure a flat surface and also to exclude O_2, which inhibits polymerization.

Davis and Ornstein (1959) and Raymond and Weintranb (1959) were the first to use PAGE instead of starch gel electrophoresis. The material used had a trade name of cyano gum 41. It consisted of a mixture of acryl amide in such proportions that a gel could be formed from 3-10 percent of n-N^1 methylene bis acryl amide. The most appropriate composition for gel formation was a mixture of acryl amide and N–N^1 methylene bisacrylamide. This composition was known as cyanogum-41.

Polyacrylamide gel electrophoresis (PAGE) is superior to other electrophoretic processes such as paper and starch gel electrophoresis due to the following reasons:

- Transparency of gel permits direct measurement of pattern by tranquitting light photometry through gels.

- Gels can be dried to thin flexible films. Rehydration to original volume is possible by immersion in water.
- Gels are chemically inert which facilities differential scanning.
- Gels are non-ionic.
- Pore sizes of the gels can be controlled by the choosing various concentrations of acryl amide and methylene bis acrylamide at the time of polymerization.
- By incorporating thedetergent SDS (sodium dodecyl sulfate) proteins can be separated and their molecular weights can also be calculated. Hence, heterogeneity of proteins can be studied using SDS–PAGE.

PAGE is commonly used to separate out various fractions of milk proteins like α lactalbumin, whey proteins, and caseins (α, β,K, & Y) under alkaline conditions (pH 8.0, 8.4 or 8.8). However, some of the components separated are not homogenous. They are better separated under acidic conditions, particularly the β-casein variants A1, A2, etc.

Protocol for Performing PAGE

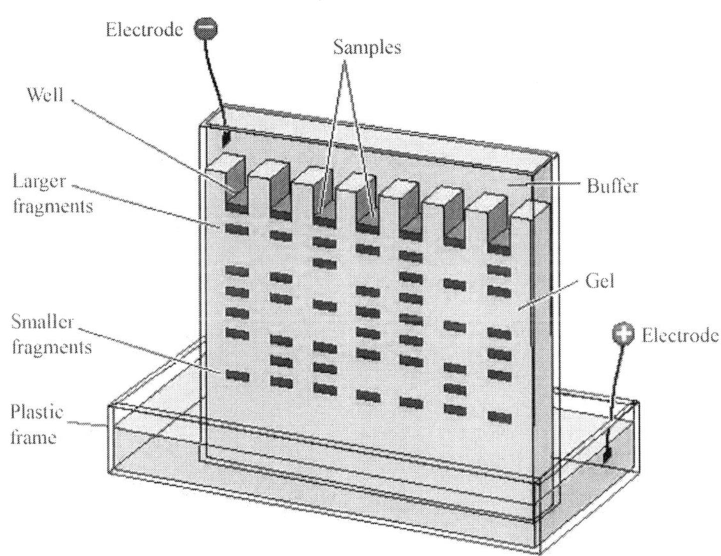

Preparation of Acryl Amide Stock Solution

Acrylamide	5.55 g
Bisacrylamide	0.15 g

Dissolve in 25 mL distilled water.

Acryl amide stock solution is also available as cyanogum 41. The ratio of acrylamide to bisacrylamide is 60:1.013, which is the most preferred one by several workers.

1. Acrylamide 22.20 g
2. Bisacrylamide 0.6 g

Dissolve in 100 mL of distilled water. Hence, a ratio of 60:1.013 is obtained.

Tris–Glycine Buffer (pH 8.3)

Buffer solution

1. 0.025 M tris 6.0 g
2. 0.192 M glycine 28.8 g

Make the volume upto 2 L. Glass-distilled water is preferred for adding to make up the volume to a liters.

Ammonium Persulphate (1% Solution)

One hundred milligrams of ammonium persulfate/10 mL for vertical gel electrophoresis 30–50 mg ammonium persulfate solution for cross-electrophoresis/10 mL Prepare the solution freshly.

Preparation of 8% Acrylamide

The following reagents are taken in prescribed amounts in a glass beaker:

1. 10.5 mL of acrylamide stock solution.
2. 18 mL of tris–glycine buffer (gel buffer).

Gel buffer contains the following ingredients

	Glycine	1.44 g
NH_2	Tris	0.3 g
C=0	urea	42 g

Make the volume 100 mL with water.

Add 0.3 ml mercaptoethanol CH_2–CH_2–oH.

3. 0.1 mL of TEMED
4. 1.5 mL of ammonium persulphate solution

Preparation of Protein Samples

1. 25 mg casein

+

2. 1 mL solubilizing buffer of the composition:

Glycine	1.44 g
Tris	0.3 g
Urea	42 g

Dilute to 100 mL with water

+

3. 0.5 mL glycerol. Glycerol is added to make the protein solution heavy so that it does not float

+

4. 0.05 mL mercaptoethanol

+

5. A drop of 0.7% (wt./vol.) amido-black in water and bromophenol blue can also be used instead of amido-black as a tracking dye.

Electrophoresis

It is done by applying 0.02–0.04 mL of protein solution and passing a current of 1 mA/sample.

Staining of Gel

Is done in 1% amido black prepared in 7% acetic acid.

1% amido black	100 mL.
7% acetic acid	100 mL.

Both are mixed together.

Destaining in 7% Acetic Acid

Glacial acetic acid	35 mL.
Distilled water	500 mL.

The gel portion containing the protein fraction will retain the dye, whereas the other portion of the gel will get decolorized by treatment with 7% acetic acid. Methanol acetic acid can be used instead of 7% acetic acid as a decolorizer to speed up the decolorization rate.

Procedure

i. Prepare all the reagents as per the procedure given in the protocol.
ii. Fill the small glass capillary tubes with 8% acryl amide gel. Rubber stoppers are inserted at the end. Fill the remaining space left at the top with water.

iii. Allow the gel to polymerize. It takes approximately 30–45 min for the gel to form.

iv. Assemble the electrophoretic apparatus. Apply the protein samples in adjusted concentration at the top of the tubes with the help of a pipette after filling the electrode buffer on both the troughs containing the +ve and –ve electrodes.

v. Pass a current of 1 mA/sample till the tracking dye comes out from the capillaries into the buffer from below. It took approximately 60–70 min for the cheese, K-casein, and whey protein samples to migrate from the –ve to the +ve pole, as indicated by the movement of the tracking dye, amido black..

vi. Disassemble the capillary tubes after stopping the current. Take out the gel column from the glass tubes with the help of a syringe and needle assembly in a trey filled with water. Transfer the gels into labeled test tubes and fill the individual test tubes containing the gel column with staining solution, i.e. 1% amido black in 7% acetic acid. Ideally, the staining is done for 1 h; however, in our case, the staining was done overnight.

vii. Remove the staining solution after the prescribed period and decolorize with 7% acetic acid in the same test tubes. Change the decolorizer at regular intervals, as indicated by the removal of color from gels. Overnight decolorization is sufficient.

viii. Examine the pattern of the obtained bands.

Starch Gel Electrophoresis

Electrophoresis is the study of the movement of charged molecules in an electric field. Electrophoretic techniques can be classified as either moving boundary electrophoresis or zone electrophoresis.

Starch gel electrophoresis is a zonal electrophoretic technique wherein a buffer-saturated solid matrix is employed as the support medium. Starch gel as a supporting medium was the first gel medium to receive attention. In these techniques, a slab gel is prepared from potato starch paste or hydrolyzed starch and is layered on a horizontal glass plate or Petri plate. The sample is placed in a well cut in the slab, and the plate is subjected to voltage. After electrophoresis, the starch slab is stained for visualization of the sample components. The enchanted resolution achieved by starch gels is probably due to the molecular sieving effect and reduced diffusion because of a more rigid network. Starch gels are still used for some protein and enzyme analysis.

Reagents

Tris-Citrate Buffer, pH 8.6

This buffer is prepared by using tris. Tris means three groups of hydroxyl methyl amine. The buffer has the following composition:

1. Tris hydroxyl methyl amine
   H2N $(CH_3 OH)_3$ = 6.897 g
2. Citric acid = 0.7881 g
3. Distilled water = 750 mL

For preparing 50 mL buffer, the quantity of chemicals mentioned above required is:

1. Tris (hydroxyl methyl) methylamine
   = 6.897 g for 750 mL
   = 0.4798 g for 50 mL
2. Citric acid = 0.05254 g
3. Distilled water = 50 mL.

Electrode Buffer/NaOH Boric Acid Buffer, pH 8.0

This buffer is used for the separation of caseins. They are negatively charged under alkaline conditions, and they separate based on the difference in charge, resulting in a better resolution.

However, some proteins, like immunoglobulin, whey proteins, etc., are denatured at the alkaline pH employed. Hence, their mobility is reduced due to denaturation and aggregation, and thus, NaOH boric acid buffer, pH 8.0, is not used for their separation. Caseins, being naturally denatured proteins, do not show such an effect, and thus the buffer is used for their separation.

The composition of the buffer is as follows:

1. Boric acid = 18.552 g
2. NaOH = 2.0 g
3. β-mercaptoethanol = 2–4 drops per 100 mL
4. Water to make the total volume to 1000 mL

The pH of the buffer is checked after preparation by pH sensitive electrode or pH strip.

The function of β-mercaptoethanol added in the buffer is to keep K-CN and β-lactoglobuline in a reduced condition so that they do not form disulphide linkages.

β-Mercaptoethanol has the following chemical structure

β α

$HS–CH_2–CH_2OH$

Amido-Black Dye Solution

It is prepared as 0.1 % concentrated solution in washing solution.

Washing Solution

Washing solution is a mixture of methanol, acetic and distilled water in 5: 1: 5 ratio volume /volume respectively.

Preparation of Starch Gel

Potato starch was used earlier for demonstration purposes. However, recently, hydrolyzed starch has been used for controlled gelling purposes. It is prepared by lysing starch by amylases to a certain required size. Better reproducibility and accuracy are obtained, and hence hydrolyzed starch is preferred to ordinary starch for research work. Starch is taken into the vacuum or suction flask.

Preparation of Starch Gel Plates

The following ingredients are used in the specified amount for preparing two starch gel plates of small (8.6 cm diameter) size.

1. Tris citrate buffer (pH 8.6) = 4.5 mL.
2. Water = 17 mL.
3. Urea = 9.7 g.
4. 2-mercaptoethanol = 0.12 mL.

Firstly, water and tris–cirate buffer is mixed in a small beaker. Half of this mixture is added to the suction flask and mixed thoroughly and quickly while heating over a flame. The other half of the mixture is added to the suction flask where the contents just start boiling. The flask is further heated, the filled solution becomes clear, transparent, and viscous. Then, urea is added to the flask and mixed. Urea prevents aggregation of caseins, while passing through the gel. Further, air bubbles are removed by suction pump. Finally, mercaptoethanol (0.12 mL) is added to the suction flask and mixed thoroughly. Then the prepared gel is added uniformly to Petri dishes and is allowed to solidify for overnight.

Preparation of Casein for Starch Gel Electrophoresis

1. Take 10 mL of milk sample in a centrifuge tube.
2. Centrifuge for 10 min. The fat plug formed at the top is removed with a spatula after keeping the tube in a cold water bath for a few minutes. Alternatively, to free fat from milk, fat plug is pierced with the help of a pointed glass rod, and milk is taken out from the side.
3. To the skim milk obtained, 1 mL of 10% acetic acid and 1 mL of 1N sodium acetate are added to bring the pH to 4.5. On shaking the contents, precipitation of casein occurs. Keep the tube as such for 10 min and then centrifuge the contents for 5 min.
4. Wash the precipitated casein 2–3 times with cold distilled water to remove excess acid. Then wash the casein two times with 5 mL of acetone. Acetone in a 5 mL quantity is added to the casein, the curd formed is broken, the contents are centrifuged, and acetone is decanted. This process is repeated. Dry the obtained casein powder on filter paper. The casein obtained can thus be used immediately.
5. For prolonged preservation, casein is washed with acetone as well as either two times each, and then it is dissolved in veronal buffer, which contains urea.

Composition of Veronal Buffer

1. Sodium baritone = 20.618 g
2. Baritone = 3.684 g
3. Distilled water = 2 L

Veronal is a trade name for barbital.

$(C_2H_5)_2$ CONH CONH CO baritone (mel. Wt 184.20)

$(C_2H_5)_2$ CONH CONH CO sodium baritone (mel. Wt 206.18)

Composition of Urea-Veronal Buffer

Two grams of urea are dissolved in 5 mL of veronal buffer, pH 8.6. 70 mg of casein in 1 mL of urea veronal buffer is dispersed or dissolved, and to this, 2–3 drops of β-mercaptoethanol are added. Small filter paper strips are taken, and they are hung on thread with the help of U-pins. Casein is applied to these filter paper strips and dried with a hair dryer. Casein layer is applied three times. And subsequently, the strips are stored.

Electrophoresis

Small bits of strips are put vertically at one end of the Petri dish. Then electrophoresis is carried out using sodium boric acid buffer, pH 8.0, at 300 Vand 3 mA current from the power supply given.

After 1½ h (as detected by the movement of tracking dye put on to the filter paper initially), the plates are removed, and amido black staining solution is added to the plates and kept for 10 min. Then the dye is replaced with the washing solution containing methanol, acetic acid, and water in the ratio of 5:1:5 until the background is clear. Subsequently, the components separated are identified.

Paper Strip Electrophoresis

Introduction

Recent advances in analytical techniques have effectively been used in obtaining and identifying the substances in high purity. Physical methods like fractional precipitation, distillation, and crystallization have been used in the separation and purification of chemical compounds. These methods worked quite successfully in many cases, but some difficulties arose when the individual components of the compounds had similar physical and chemical properties. For example, the fractional distillation method could be safely used in the case of a mixture of liquids that have a good range of boiling point differences. However, this method could not be used in the case of liquids as well as rare gases where the boiling points of individual components are very close to each other. Likewise, after treating the biological material by the usual classical methods, one is usually left with a number of compounds, such as a mixture of amine acids, which are similar in properties to each other. In such cases, extreme temperatures, pH, organic solvents, and the use of oxidizing and reducing agents are avoided, as these may irreversibly change the structure of the molecules and destroy their biological activity.

Such complicated separations were achieved by the techniques of chromatography and paper electrophoresis. These methods resolve the individual components under relatively mild conditions and utilize differences in the basic physical properties of the individual molecules. Such as their mass, size, shape, charge, and absorption properties. The paper electrophoresis method has an edge over paper chromatographic methods, as in the latter, substances with low distribution coefficients are not properly separated. Such difficulties are not faced with the techniques of paper electrophoresis, which involve the migration of charged substances under the influence of electric current. Paper electrophoresis is an incomplete form of electrolysis that is widely used in the separation of biological material and many other rare and costly substances in compounds or a mixture of compounds where some ionize and others do not. The degree of separation can be effectively determined by the use of paper electrophoresis.

Technique

The paper to be used for electrophoresis is first wetted with an electrolytic solution, which is normally a buffer solution. Baritone buffer at pH 8.6 is used. The wet paper is placed in a vessel made up of anodic and catholic compartments in such a manner that the two ends dip inside the solution. Test solutions are put on the middle portion of the paper, which is dried by pressing in between filter papers, and the test solutions are put at the middle portion of the order of 2–10 V/cm. Is then passed through the solution. Buffer solutions serve the purpose of conductors, and the wet paper acts as a connecting bridge between the two compartments. The substances move either to the cathode or anode, depending on the nature of the charge that they possess, as shown in the figure.

The movement of the substances is expressed by eletrophoretic mobility (μ) which is defined as"

Distance moved/unit time

μ = electrical field strength, cm/s

Whose unit is V/cm.

$= cm^2\ V\ s^{-1}$

The movement of the substance depends on many factors which are listed below:

1. Size and nature of charge.
2. Environment factors like the concentration of electrolyte, ionic strength, dielectric properties, chemical properties, temperature, viscosity, etc.
3. pH effect,
4. Diffusion, and
5. electro-osmosis

Various procedures commonly used for performing paper electrophoresis include:

1. Sandwich.
2. Ridge pole.
3. Solvent immersion, and
4. Horizontal strip techniques.

Paper electrophoresis has been widely used in the field of large molecules such as proteins, enzymes, nucleic acids, etc.; now it is also used in the separation of peptides, nucleotides, amino acids, and abnormal hemoglobin.

Reagents

Composition

Veronal Buffer, pH 8.6 Containing 10% Urea

Sodium baritone 20.618 g

$(C_2H_5)_2$ CONH CONa CO

Barlitone

$(C_2H_5)_2$ CONH CONa CO 3.684 g

Distilled water 2.0 L

Add 10 mg urea/100 mL veronal buffer.

Bromophenol Blue Solution, 0.1% Preparation

100 mg bromophenol blue,

pH 2.4–4.6

+

100 mL ethanol

+

Saturated mercurous chloride

(Hg_2Cl_2)

Two Percentage Casein

Add 5 mL of veronal buffer with 10% urea to 100 mg of dry (pH 8.6) casein.

Acetic Acid, 0.5% Solution

Procedure

1. Cut strips of 37 × 4 cm size from 3 mm sheets of Whatman filter paper No. 31. Soak these strips in veronal buffer pH 8.6, containing 10% urea. Blot the excess buffer with blotting paper and place the strips on the electrophoretic tank horizontally. Apply samples at a distance of 8 cm from the cathode end. 2% casein in 0.1 mL quantity is spread in the central portion of the starting line of paper strips.

2. Run these samples for 6 h at room temperature, keeping the voltage constant at 220 in veronal buffer, pH 8.6. Six strips can be run at a time. Keep the current flow between 15 and 17 mA.
3. After completion of the run, dry the strips at 37 °C in an oven.
4. Stain the strips with 0.1% solution of bromophenol by soaking them in it for 30 min.
5. Wash the excess dye three times with 0.5% acetic acid subsequent to staining.
6. Dry the stained stripes in an oven.
7. Identify the separated components in the test sample by comparison with the pattern obtained for pure fractions like α_s and K-casein prepared from whole milk.

Paper strip electrophoresis has been used for the resolution of both acid and micellar casein into α and β components by the procedure of Aschaffenburg.

Chromatography

Introduction

Michal Tswett, a Russian botanist, initially developed chromatography, a relatively modern method, in Warsaw in 1906 to separate colored substances into their component parts. Since then, the method has undergone significant changes, leading to the current use of several forms of chromatography to test the purity of the constituents and separate nearly any mixture—colored or colorless—into its constituent parts. Chromatography is the Greek word for "colour writing" (chrome, colour) and "graphy, writing."

Chromatography may be defined as a separation technique of the components of mixtures by a continuous distribution of the components between two phases, one of which is moving and the other is static. The systems associated with this definition are as follows:

i. A solid stationary phase with a liquid or gaseous mobile phase and
ii. A liquid stationary phase with a liquid or gaseous mobile phase.

The system (i) gives rise to adsorption chromatography whiles the system (ii) to partition chromatography.

Chromatography differs from other methods of separation in respect of materials, equipment, and technique. It is basically dependent on two principles *viz.* adsorption and partition.

Classification

Either a liquid or a gas could be the moving phase. Depending on the characteristics of the moving and fixed phases the various types of chromatography are as follows:

Adsorption Chromatography

It refers to the use of a stationary phase or support, such as an ion exchange resin, that has a finite number of relatively specific binding sites for solute molecules. It is based on the differences in the adsorption coefficients. In this the fixed phase is a solid, e.g. alumina, magnesium oxides, silica gel etc. the solutes are adsorbed in different parts of the adsorbent column. The adsorbed components are then eluted by passing suitable solvents through the column. Adsorption techniques, represented by ion-exchange chromatography, are most effective when applied to the separation of macromolecules including proteins and nucleic acids.

Partition Chromatography

It is the distribution of a solute between two liquid phases, one fixed and the other mobile. It operates by mechanism analogous to counter–current distribution. Partition chromatography may involve direct extraction using two liquids, or it may use a liquid immobilized on a solid support as in the case of paper, thin- layer, and gas –liquid chromatography. For partition chromatography, the stationary phase consists of inert solid particles coated with liquid absorbent. The distribution of solutes between the two phases is based primarily on solubility differences. The distribution may be quantified by using the partition coefficient, K_D.

Concentration of solute mobile phase

K_D = concentration of solute in stationary phase.

Also, the similarity or identity of unknown compound as compared to standard can be established by color development or retardation value (R_f value).

$$R_f \text{ value} = \frac{\text{Distance moved by the solute}}{\text{distance moved by the solvents}}$$

The R_f value would be constant for a similar compound of similar structure, and hence the identity of test sample can be established by finding out its R_f value.

A partition system is manipulated by changing the nature of the two liquid phases, usually by combination of solvents or pH adjustment of buffers. Usually, the more polar of the two liquids is held stationary on the inert solvent, which is used to elute the sample components (normal-phase chromatography). Reversal of this arrangement, using a non-polar stationary phase and a polar mobile phase, is known as reversed-phase chromatography. Polar hydrophilic substances, such as amino acids, carbohydrates, and water-soluble plant pigments, are separable by normal-phase partition chromatography. Lipophilic compounds, such as lipids and fat-soluble pigments, may be resolved with reversed-phase systems.

Partition chromatography has been widely used for the separation and identification of amino acids, carbohydrates, and fatty acids.

Gas Chromatography

The moving phase is a mixture of gases; it is called gas chromatography.

Depending on whether the compound of interest has to be separated or identified, chromatographic techniques are classified as preparative or analytical.

A preparative procedure is one that can be applied to the purification of a relatively large amount of biological material. The purpose of such an experiment would be to obtain purified material for further characterization and study.

Analytical procedures are used most often for the determination of the purity of a biological sample; however, they may be used to evaluate any physical, chemical, or biological characteristic of a biomolecule or biological system. Here, the layer of gel or paper used is kept thin, e.g. Whatman filter paper No. 1 is used.

On the basis of solute material used, chromatography has been classified as

1. Liquid–liquid chromatography
2. Gas–liquid chromatography
3. Solid–liquid chromatography and so on.

Depending on the support material used, chromatography has been classified as paper, thin layer, column chromatography, gel permeation chromatography, etc. The columns used in column chromatography may be calcium carbonate, sephadex, resins, DEAE cellulose, hydroxyl methyl cellulose, alumina, etc.

Depending on the method of development used, chromatography can be classified as ascending or descending chromatography.

If chromatography runs in one direction, then it is called unidirectional chromatography, and if it runs in two directions, it is called two-dimensional chromatography.

Depending upon the type of spot material, chromatography can be classified as circular and paper strip.

Thus, there are numerous chromatographic techniques based on the principle of either adsorption or elution.

Chromatographic separations in practice may be done in any one of the three forms: column chromatography, in which the stationary phase is packed into glass or metal columns; thin layer chromatography, in which the stationary phase is thinly coated onto glass, plastic, or foil plates; and paper chromatography, in which the stationary phase is supported by the cellulose fibers of a paper sheet. Each of these three forms of chromatography has its own specific advantages, applications, and mode of operation.

Column chromatography is the method used for the isolation and preparation of compounds, whereas paper or thin layer chromatographic techniques are used for analytical work.

Column Chromatography

Column chromatography is one of the frequently used method in biochemical work for the separation of compounds.

Columns:

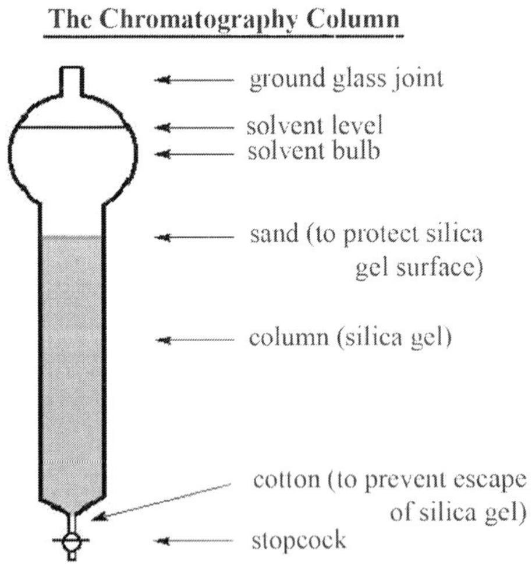

Glass is typically used for chromatography columns. Long columns typically provide good component resolution, but wide columns work better when handling large amounts of material. The diagram above depicts the key components of a chromatographic column.

Preparation of the Material

Chromatographic separations require the equilibration of a wide range of materials with the solvent column preparation. Furthermore, pre-treatment is frequently necessary. For instance, some gel filtration materials must be swollen, adsorbents must be "activated" by heating or acid treatment, and washing is necessary to obtain ion exchange resins in the necessary ionized form.

The material is allowed to settle during the solvent equilibrium, and any fine particles that remain in suspension are extracted by decantation. If this isn't done, these tiny particles will clog the column and significantly lower the solvent flow rate.

The Pouring of the Column

To prevent air bubbles from becoming trapped in the column, the chromatography column is packed with material by slowly adding a slurry of the material in the solvents until it is about one-third full. This is done by carefully pouring the material down a glass, as shown. Excess solvent is allowed to run off and the suspension to settle. Until the column reaches the desired height, these steps are repeated. After that, a full solvent wash is performed on the column, with the liquid level maintained slightly above the material's surface.

Application of the Sample

Before being loaded onto the column, the sample is either dialyzed against the eluting buffer or dissolved in the solvent. The concentrated sample is typically pipetted carefully onto the surface during class experiments, and the tap is then opened until the top of the column is slightly below the meniscus. A pressure reservoir provides a steady head of liquid to be maintained at the top of the column, and the solvent reservoir is connected to it.

Elution

The next step involves eluting the materials with the proper solvent in order to remove them from the column in order.

The Collection and Analysis of Fractions

Either manually or using a fraction collector, the column's effluent is collected into several test tubes. The compounds under investigation are then checked for in each fraction, and an elution profile is created by comparing the amount eluted with the effluent volume.

Paper Chromatography

Principle

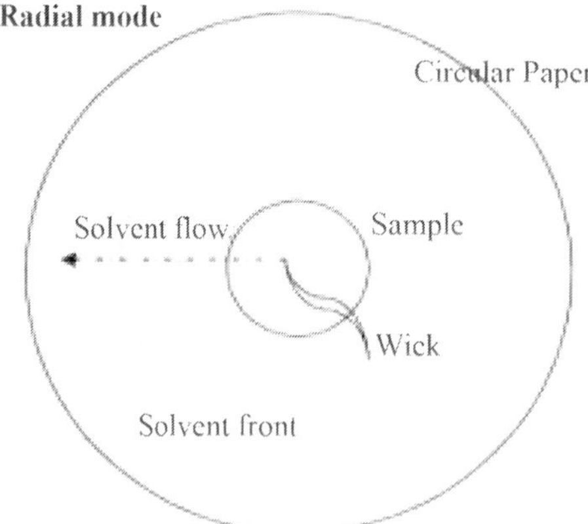

Cellulose in the form of paper sheets makes an ideal support medium where water is adsorbed between the cellulose fibers and forms a stationary hydrophilic phase.

The mixture is spotted on the paper, dried, and the chromatogram developed by allowing the solvent to flow along the sheet. The solvent front is marked, and, after drying the paper, the portions of the compounds present in the mixture are visualized by a suitable staining reaction. The ratio of the distance moved by a compound to that moved by the solvent is known as the RF value and is more or less constant for a particular compound, solvent system, and paper under carefully controlled conditions of solute concentration, temperature, and pH.

Paper

Whatman No.1 is the most frequently used paper for analytical purposes. Whatman No.3 is a thick paper and is best employed for separating large quantities of material. The resolution is, however, inferior to Whatman No.1. For a rapid separation, Whatman Nos. 4 and 5 are convenient, although the spots are less well defined. The paper may be impregnated with a buffer solution before use or chemically modified by acetylation. Ion exchange papers are also available commercially for the separation of lipids and similar hydrophilic molecules; silica-impregnated papers are also available commercially.

Solvents

This choice, like that of the paper, is largely empirical and will depend on the mixture investigated. The pH may also be important in a particular separation, and many solvents contain acetic acid or ammonia to create a strongly acidic or basic environment.

Ascending Chromatography

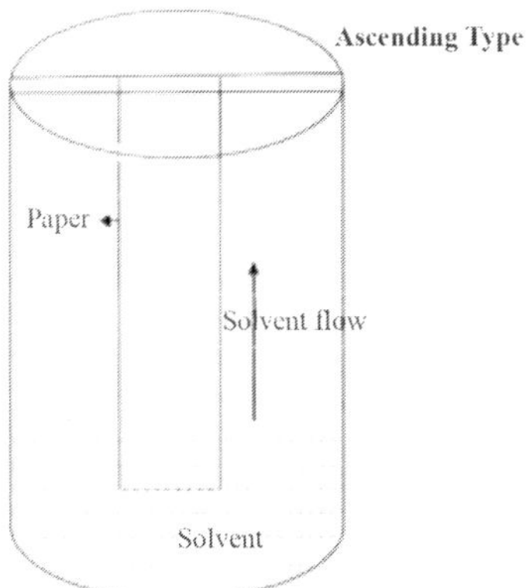

The sheet of paper is supported on a frame with the bottom edge in contact with a trough filled with solvent. Alternatively, the paper can be rolled into a cylinder, fastened with a paper clip, and stood in the solvent. The arrangement is contained in an air-tight tank lined with paper saturated with the solvent to provide a constant atmosphere, and separations are carried out in a constant temperature room.

Descending Chromatography

The method is convenient for compounds which have similar R_f values since the solvents drips off the bottom of the paper, thus giving a wider separation. In this case, the R_f values cannot be measured with a standard reference compound such as glucose, for example in the case of sugars.

Two-dimensional Chromatography

The mixture is separated in the first-solvent, which should be volatile.Then after drying, the paper is turned to 90° and separation is carried out in the second solvent. After location, a map is obtained and compounds developed under the same conditions.

Detection of Spots

Most compounds are colorless and are visualized by specific reagents. The location reagent is applied by spraying the paper or rapidly dipping it in a solution of the reagent in a volatile solvent. Viewing under ultraviolet light is also useful since some compounds that absorb strongly show up as dark spots against the fluorescent background of the paper. Other compounds show a characteristic fluorescence under ultraviolet light.

Thin Layer Chromatography

Introduction

Thin-layer chromatography is adsorption chromatography performed on open layers of adsorbent materials supported on glass plates. It is a separation method in which uniform, thin layers of sorbent or selected media are used as a carrier medium. The sorbent is applied to a backing as a coating to obtain a stable layer of suitable size. The most common supports, such as plastic sheets and aluminum foil, are also used. The sorbents most commonly used are silica gel, alumina, Kieselguhr (diatomaceous earth), and cellulose.

The term sorbent is used in a general sense to include layer materials that may be used for either adsorption or partition chromatography. The standard size for TLC plates is 20 × 20 cm. For most separations, the mobile phase is allowed to travel on the layer for a distance of 15 cm. Other plate sizes used are 5 × 20 cm 10 × 20 cm, and 20 × 40 cm. "Micro" plates have been made from microscope slides.

In practice, a sample to be separated is applied on the layer 1–2 cm from one end of the plate. The further edge of the application is called the starting point or origin. Separation is achieved by passing a solvent, the mobile phase, through the layer. The layer, with the sample zones at the bottom, is placed ata slight angle from the vertical in a closed tank containing a small amount of the mobile phase.

The nature and chemical composition of the mobile phase are determined by the type of substance to be separated and the type of sorbent to be used for the separation. The composition of a mobile phase can be as simple as a single, pure solvent (such as benzene used to separate dyes on alumina) or as complex as a three- or four-component mixture containing definite proportions of chemically different substances, such as a 1:1:1:1 solution of n-butanolethyl–acetate–acetic acid–water used to separate amino acids on silica gel.

Capillary action causes the mobile phase to travel through the medium in a process called development. Ascending development is the most common, but horizontal, descending, and centrifugal methods have also been reported. After the plate is dried, the separated spots can be visualized in several ways such as viewing under an ultraviolet light or spraying with one of a wide variety of reagents. The R_f value is a convenient way to express the position of a substance on a developed chromatogram. It is calculated as the ratio:

$$R_f \text{ value} = \frac{\text{Distance of compound from origin}}{\text{distance of solvent from origin}}$$

R_f value ranges between 0 and 0.999, and without units. Distance is measured to the center of the sample zone or spot.

According to the types of compounds being investigated, the separations on the plate may then be evaluated by a number of instrumental techniques. If the substances are radioactive or suspected to be so, the plate can be examined by a radioisotope scanner to locate them. Under proper conditions, it is also possible to measure both the area of a spot and its density using a photodensitometer. These characteristics are related to the quantity of a given substance in the sample and can thereby be used to assess the amount of the substance in the sample. A quantitative calibration or reference curve should be established beforehand using known amounts of the substance to be quantitated.

Thin layer chromatography (TLC) has found wide applicability in separation and identification of both organic and inorganic substances.

The main advantage of TLC lies in the fact that the technique can carry out the separation and identification of unknown substances in very small amounts and in a very short time. None of the other methods can match it in this regard. Even paper chromatography, which has earned the reputation of carrying out separation in very small amounts, has been exceeded by this method. For example, TLC can carry out separation or identification in still smaller amounts and in much less time than is taken by paper chromatography.

TLC has been used in separation and identification of a large variety of organic compounds: alkaloids, terpenes, vitamins, hormones, dyes, drugs, sugars, fat, pigments, alcohols, aldehydes, acids, and many others. TLC has also been used in the analysis and identification of inorganic compounds.

Techniques and Materials

Absorbents and applicators are the two essential items of the TLC techniques, although other accessories like glass plates, aligning trays, activating ovens, developing chambers, etc. play their individual roles. The efficiency of the former two plays a very important part in the success of the technique.

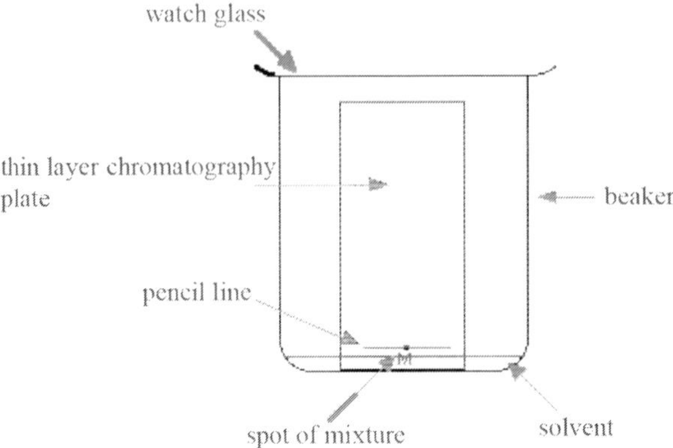

Applicator

There are various types of applicators available for spreading the adsorbent in the form of slurry over the glass plates that are used in chromatography. Adjustable applicators are available for spreading layers of various degrees of thickness ranging from 250 μ to 2 mm over glass plates.

Adsorbents

The character of the adsorbent is of the utmost importance in this technique. Silica gel and alumina are mostly used and have proved to be quite suitable. They are used along with some binding materials, like plaster of Paris, to fix the thin and porous layer on the glass plates. In addition to this, Kiesel gel has also been used for making thin layers. Cellulose powders with and without binding materials are also in use. Various types of cellulose powder, including ion-exchange cellulose powder, are widely used.

Preparation of Thin Layers

The adsorbents are thoroughly mixed with the requisite quantities of water or mixtures of water and alcohol with the help of a water mechanicalstirrer for about 2 min to form a homogenous slurry. Then, immediately, it is spread over well-cleaned glass plates of standard size (20 × 20 cm or 20 × 5 cm) arranged in a row on an aligning tray with the help of the applicator. The applicator is first adjusted according to the thickness of the layer, after which the slurry is put inside the applicator, and the latter is drawn by hand over the glass plates arranged on the aligned tray. The coated glass plates are then allowed to dry at room temperature for 10 min to 2 h, during which the linder present in the adsorbent sets the adsorbent firmly over the glass plates.

After air drying, the coated glass plates are activated by drying in an oven at a suitable temperature, depending on the adsorbent. The heating period similarly varies. After activation, the plates are stored in vacuum desiccators until they are used. If the plates are not stored in desiccators and exposed to air or kept for a long time without being used, the efficiency of the plates decreases.

The samples to be analyzed are applied in the same way as in paper chromatography. Usually, a small amount, e.g. 4 μL of the solution of the substance or a mixture in a micro-pipette is applied to the starting point near one of the edges (narrow edge) of the chromatoplates in such a manner that the drop becomes particable.5–10 μg are sufficient for one operation. Substances are usually applied about 2 cm from one side of the plate.

The plates are then put in the chromatography chamber, which is usually much smaller than the paper chromatography chamber. The chamber should be sufficiently saturated with solvent vapor ahead of the actual operation. Its walls should be lined up with filter paper soaked with the solvent to maintain steady saturation inside the chamber. The solvent is put inside the chamber to a thickness of 1 cm. The spotted chromatoplates are developed by ascending techniques. The plates are held vertically inside the chamber. They can also be developed by the descending technique with certain modifications to the chamber. The development takes much less time than in the case of paper chromatography. When the development is complete, the plates are taken out and allowed to dry in the air inside a clean, dust-free chamber at room temperature. Then the developed plates are activated by keeping them in an oven adjusted to a specified temperature (100–250 °C).

The detection of the movement of spots and the determination of R_f values are similar to those in paper chromatography. Detection of spots can be made by spraying the chromatogram with visualizing reagents (e.g., ninhydrin) or, in some cases, by seeing the chromatogram in UV light.

Partition and ion exchange TLC can be performed in the same way. Two-dimensional thin-layer chromatography has also been achieved with success by developing the single plate in two directions at right angles to each other, utilizing two different solvent systems. Thin-layer chromatography has shown good success in the separation of mixtures of various types of compounds.

Separation of Milk Fat Components by TLC

Procedure

Preparation of TLC Plate

Silica gel G was taken in a quantity of 4 g, 11 mL of distilled water was added, and a slurry was made with pastel and a grinder. Further, the slurry was poured over glass plates of 10 × 20 cm size uniformly and spread evenly with the help of a glass rod or special apparatus available for the purpose. The plates were then air dried and water evaporated. The plates, after completion of drying, were activated at 110 °C for 20 min. During activation, much of the moisture gels evaporate and the silica gel gets properly charged. These plates, after activation, were spotted with the ghee sample.

Preparation of Sample

11 mL of fat sample was solubilized in 1 mL of hexane and spotted on TLC plates with the help of glass capillary. Spotting is done in a manner that the gel is not pulled out. The amount of the sample to be analyzed should also be standardized. If the area of spotting is kept too large, overlapping of components occur and a poor resolution between molecules which retards their migration.

Development

After spotting the standard as well as the unknown, plates were put in a chromatographic chamber containing the solvent system. To aid saturation of the developing chamber atmosphere, the inner walls are normally lined with blotting paper or a thick chromatography paper such as Whatman 3. This is conveniently done by taking a single large sheet of paper and placing it into the dry tank so that the back and ends are lined. The front is left open so the plate may be observed. The paper should come to within 1 cm of the top of the tank. The chromatography was allowed to proceed for a given period of time, and then the plates were removed from the other end, air dried, and taken for color development.This can be done using any one of the two solvent systems A and B.

Composition of solvent A

Hexane, diethyl ether and glacial acetic acid in the ratio of 70:30:1.

Composition of solvent B

Petroleum ether (boiling point 250 °C), diethyl ether and glacial acetic acid in the ratio of 80:20:1.

Detection of the Separated Pattern

The plates were air dried and sprayed with 50% H_2SO_4 and heated at 110 °C for 30 min. The components were tentatively identified according to their mobility of various lipid components as long chain triglycerides, > short chain triglycerides, > free fatty acids, > diglycerides, > cholesterol > 1.2 diglycerides > 2 monoglycerides > phospholipids.

Separation of Pigments

Procedure

In a mortar, place 5 mL of petroleum ether + ether and a few green leaves. Crush the leaves with pastel and transfer the extract to a separating funnel by means of a pipette. Swirl the extract with an equalvolume of water. Discard the lower aqueous layer. Repeat washing two times with water. Discard the lower aqueous layer. Take the petroleum ether layer to a small aluminum flask anddry over 2 g of sodium sulfate,or if concentrated gel is used, by a stream of dry nitrogen. Place a small spot with the help of a capillary tube on a 10 cm TLC plate. Allow the spot to dry. Use acetone as developing solvent. Develop the chromatogram using a wide-mouth bottle or a conical flask. A folded piece of filter paper is placed in a bottle to moisten or saturate the atmosphere with water. Dry the plates again, and it will be possible to observe as many as eight colored spots. In order to decrease the $R_f$value, the following spots will be developed of different separated pigment compounds:.

Carotenes → two orange-colored spots
Chlorophyll A → one blue–green colored spot
Chlorophyll B → one green spot
Xanthophylls → four yellow-colored spots

Pates used for performing chromatography are either silica gel or alumina of 0.25 mm thickness.

Affinity Chromatography

Introduction

Affinity chromatography is a unique separation technique that does not rely on differences in the physical properties of the molecules to be separated. Instead, it exploits the unique propertiesof extremely specific biological interactions to achieve separation and purification. As a consequence, affinity chromatography is theoretically capable of giving absolute purification, even from complex mixtures, in a single process. The technique was originally developed for the purification of enzymes but it has been extended to nucleotides, nucleic acids, immunoglobulin, membrane receptors, and even whole cells and cell fragments.

The technique requires that the material to be isolated is capable of reversible binding to a specific ligand which is attached to an insoluble matrix.

$$\underset{\text{Macro-Molecule}}{M} + \underset{\text{ligand (attached to matrix)}}{L} \underset{k-1}{\overset{k+1}{\rightleftharpoons}} \text{ML complex}$$

Under the correct experimental conditions, when a complex mixture containing the specific compound to be purified is added to the insolubilized ligand, generally contained in a conventional chromatography column, only that compound will bind to the ligand. All other compounds can therefore be washed away, and the compound is subsequently recovered by displacement from the ligand.

Practical Procedure

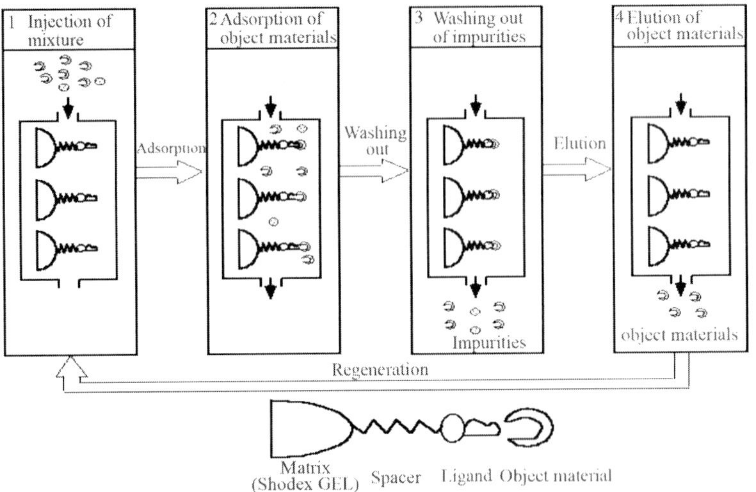

Under the correct experimental conditions, a complex mixture containing the specific compound to be purified is added to the insolubilized ligand, generally contained in a conventional chromatography cola. The procedure for affinity chromatography is similar to that used in other forms of liquid chromatography. The ligand-treated matrix is placed into the column in the normal way for the particular type of support. A buffer is used, which will encourage the binding macromolecule to be strongly bound to the ligand. The buffer generally has a high ionic strength to minimize the non-specific adsorption of polyelectrolyte ligands. The buffer must also contain any co-factors, such as metal ions, necessary for ligand-macromolecule interaction. Once the sample has been applied and the macromolecule bound, the column is eluted with more buffers to remove non-specifically bound contaminants; only that compound will bind to the ligand. All other compounds can therefore be washed away, and the compound is subsequently recovered by displacement from the ligand. The purified compound is finally recovered by either specific or non-specific elution. Non-specific elution may be achieved by a change in either pH or ionic strength. pH shift elution using dilute acetic acid or ammonium hydroxide results from a change in the DNA by a similar procedure. The method has also been used to isolate whole genes by hybridizing practically single-stranded DNA with complementary mercurated mRNA, which is then reacted with a thiolated matrix.

Gas Liquid Chromatography (GLC)

Introduction

Gas liquid chromatography is also known as vapour phase chromatography, gas–liquid partition chromatography, gas-partition chromatography, gas-chromatography, vapour fractometry and by several similar designations.

This technique, which is based upon the distribution of compounds between a liquid and a gas phase, is a widely used method for the qualitative and quantitative analysis of a large number of compounds. It has high sensitivity, reproducibility, and speed of resolution. It is proved to be the most valuable technique for the separation of compounds of relatively low polarity. A stationary phase of 'liquid' material such as silicone grease is supported on an inert granular solid. This material is packed into a narrow-coiled glass or steel column 1–3 m long and 2–4 mm internal diameter through which an inert gas (the mobile phase) such as nitrogen or argon is passed. The column is maintained in an oven at an elevated temperature which volatilizes the compounds to be analysed. The basis for the separation of compound being analysed is the difference in the partition coefficient of the volatilized compounds as they are carried through the column by the gas. As the compounds leave the column they pass through a detector. This is linked via an amplifier to a chart recorder, which records a peak as a compound passes through the detector.

Apparatus and Working

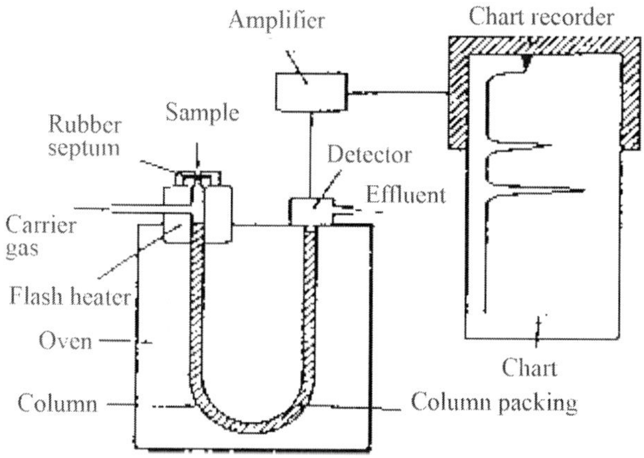

Diagram shows the essential parts of the apparatus. Helium gas is introduced at a constant rate of flow through value, passing through detector then through the column (stationary phase) filled with silicon or other non-volatile liquid, then through lower part of detector, then through gas flow meter M where exact rate of flow of gas is measured. The sample to be analyzed is dissolved in carrier solvent and injected at the top of the column with STD. The detector is a thermal conductivity cell called a double katharometer. The principle of this detector is Wheatstone bridge. Heat is condensed away from a hot body by passing a gas over hot body. The rate of heat removal will depend on the nature of the gas, other factors being constant. Thus 1/5th the heat conductivity of air whereas H_2 gas is seven times more effective than air.

Solid Support

Since this is used to provide a supporting surface on which the film of stationary phase is coated, it is important that the support should be inert to the sample. The most commonly used support is celite (diatomaceous silica) which because of the problem of support/ sample interaction, is often treated so that hydroxyl groups which occur in celite are modified. The support particles have an even size which, for the majority of practical applications it is 60–80, 80–100, or100–120 meshes.

Stationary Phase

The requirements for any stationary phase are that it must be non-volatile and thermally stable at the temperature used for analysis. Commonly used stationary phases include the polyethylene glycols, methyl phenyl- and methylvinylsilicone gums (so called OV phases), Apiezon L, esters of adipic, succinic and phthalic acids, and squalene.

Preparation and Application of Sample

The majority of non-and low polar compounds are directly amenable to GLC, but other compounds possessing such polar groups as –OH, –NH2, –COOH are generally retained on the column for excessive periods of time if they are applied directly. This excess time is inevitably accompanied by poor resolution and peak tailing. This problem can be overcome by derivatisation of these polar groups. Methylation, silanisation, and trifluoromethyl-silanisation are common derivatisation methods for fatty acids, carbohydrates, and amino acids.

The sample for chromatography is dissolved in a suitable solvent such as ether, heptanes or methanol. The sample is injected into the column using a micro-syringe through a septum in the injection port which is attached to the top of the column. Normally between 0.1 and 10 mm^3 of solution is injected. It is a common practice to maintain the injection region of the column at a slightly higher temperature than the column itself. This helps to ensure rapid and complete volatilization of the sample. Sample injection is automated in many commercial instruments.

Separation Conditions

Nitrogen, helium, and oxygen are the three most commonly used carrier gases. They are passed through the column at a flow rate of 40–80 cm^3 min^{-1}; the column temperature must be within the working range of the particular stationary phase and is chosen to give a balance between peak retention time and resolution. In isothermal analysis a constant temperature is employed. In the separation of compounds of weight it may be advantageous to gradually increase the temperature. This is referred to as temperature programming.

Detection Systems

By far the most widely used detector is the flame ionizing detector (FID). It responds to almost all organic compounds, can detect as low as 1 ng and has a wide linear response range. A mixture of hydrogen and air is introduced into the detector to give a flame, the jet of which forms one electrode, whilst the other electrode is brass or platinum wire mounted near the tip of the flame. When the sample components emerge from the column they are ionized in the flame, resulting in an increased signal being passed to the recorder. The carrier gas passing through the column and the detector gives a small background signal, which can be offset electronically to give a baseline.

The nitrogen–phosphorous detector (NPD), which is also called a thermionic detector, is similar in design to a FID, but has a sodium salt fused onto the electrode system or a burner tip embedded in a ceramic tube containing a sodium salt or a rubidium chloride tip.

The electron capture detector (CED) responds only to substances which capture electrons, particularly halogen-containing compounds. This detector is, therefore, particularly used in the analysis of polychlorinated compounds.

The volatile solvent used to introduce the test sample gives rise to a solvent peak at the beginning of the chromatogram.

The three main forms of detector respond to this solvent with varying sensitivity affecting the detection and resolution of rapidly eluting solutes. In cases where authentic samples of the test compounds are not available for calibration purposes or in cases where the identity of the compounds is not known, the detector may be replaced by a mass spectrometer. More recently, GLC has been linked to other types of detector, including an infra-red spectrophotometer and to a nuclear magnetic resonance spectrophotometer, the resulting spectra aiding in the identification of unknown compounds.

Applications

Until the recent developments in HPLC, GLC was probably the commonly used form of chromatography. Its use nowadays is confined to volatile, non-polar compounds which do not need derivatisation where compounds are characterized by their retention time or preferably by their relative retention time to a standard reference compound. In the analysis of compounds which form a homologous series, for example the methyl esters of the saturated fatty acids, there is a linear relationship between the logarithm of the retention time and the number of carbon atoms. This can be exploited, for example, to identify an unknown fatty acid ester in a fat hydrolysate. A widely used system for quantitative analysis is the retention index(RI) which is based on the retention of a compound relative to n- alkanes. The compound is chromatographed with a number of n-alkanes and a semi- logarithmic plot constructed of retention time against carbon number. Each n-alkane is assigned an RI of 100 times the number of carbon atoms it contains (pentane therefore has an RI of 500) allowing the RI for the compound to be calculated. Many commercially available GLC systems with data processing facilities have the capacity to calculate RI values automatically.

High Performance (Pressure) Liquid Chromatography (HPLC)

Principle

Originally, HPLC was referred to as high pressure liquid chromatography but nowadays the term high performance liquid chromatography is preferred since it better describes the characteristics of the chromatography and avoids creating the impression that high pressure are an inevitable pre-requisite for high performance. The components of a HPLC system are shown in the following diagram:

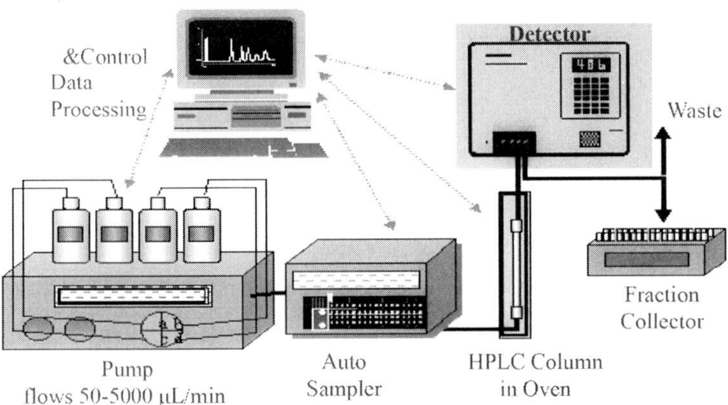

The resolving power of a chromatographic column increases with column length and the number of theoretical plates per unit length, although there are limits to the length of a column due to the problem of peak broadening. As the number of theoretical plates is related to the surface area of the stationary phase it follows that the smaller the particle size of the stationary phase, the better the resolution. Unfortunately, the smaller the particle size, the greater the resistance to eluent flow. All of the forms of column chromatography so far discussed rely on gravity or low-pressure pumping systems for the supply of eluent to the column. The consequences is that the flow rates achieved are relatively low and this gives greater time for band broadening by simple diffusion phenomena. The use of faster flow rates is not possible because it creates a back-pressure which is sufficient to damage the matrix structure of the stationary phase, thereby actually reducing eluent flow and impairing resolution. In the past decade there has been a dramatic development in column chromatography technology which has resulted in the availability of new smaller particles size stationary phases which can withstand these pressures and of pumping systems which can give reliable flow rates. These developments, which have occurred in adsorption, partition, ion- exchange, exclusion, and affinity chromatography, have resulted in faster and better resolution and explain why HPLC has emerged as the most popular, powerful, and versatile form of chromatography.

Column Packing

Three forms of column packing material are available based on a rigid solid (as opposed to gel) structure. These are:

i. Microporous supports where micropores ramify through the particles which are generally 5–10 μm in diameter;

ii. Pellicular (superficially porous) supports where porous particles are coated onto an inert solid core such as a glass bead of about 40 μm in diameter:

iii. Bonded phases where the stationary phase is chemically bonded onto an inert support.

For adsorption chromatography, adsorbents such as silica or alumina are available as microporous or pellicular forms with a range of particle size. Pellicular systems generally have a high efficiency but low sample capacity, and therefore microporous supports are preferred where applicable. All forms of HPLC column packing are characterized by their regular spherical shape which distinguishes them from conventional materials. These small spheres pack most efficiently and give good flow properties.

In liquid–liquid partition systems, the stationary phase may be coated on the inert support. Both microporous and pellicular supports are used for supporting the liquid phase. One disadvantage of supports coated with liquid phase is that the developing solvent may gradually wash off the liquid phase with repeated use. To overcome this problem bonded phases have been developed where the liquid phase has been covalently bonded to the supporting material which may be silica or a silicone polymer. The silicone polymer bonded phases have the particular advantage that as well as not being eluted by the developing solvent, they are chemically, hydrolytically and thermally stable. In normal phase liquid–liquid chromatography, the stationary phase is a polar compound such as alloys nitride or allylamine derivatives and the mobile phase a non-polar solvent such as hexane. For reverse-phase chromatography, the stationary phase is non-polar compound such as a C_8 or C_{18} hydrocarbon and the mobile phase a polar solvent such as water/acetonitrile or water/methanol mixtures.

Many different types of ion exchangers are available of which the cross-linked microporous polystyrene resins are widely used. Pellicular resin forms are also available, as are bonded phase exchangers covalently bonded to a cross-linked silicone network. These resins are classed as hard and readily withstand the pressures required during analysis.

The stationary phases for exclusion separations are generally porous silica, beads, polystyrene, or polyvinyl acetate the eluting solvent is an organic system, and the beads are available in a range of pere sizes. Semi rigid gels such as sephadex or bio-gel P and non-rigid gels such as sepharose and bio-gel A are only of limited use in HPLC since they can withstand only low pressure. The supports for affinity separations are similar to those for exclusion separations. The spacer arm and ligand are attached to these supports by similar chemical means to those used in conventional low pressure affinity chromatography.

The Column

The columns used for HPLC are generally made of stainless steel and are manufactured so that they can withstand pressure of up to 5.5×10^7 Pa (8000 psi). Straight column of 20–50 cm in length and 1–4 mm in diameter isgenerally used though smaller capillary columns are available. The best columns are precision bored with an internal mirror finish which allows efficient packing of the column. Porous plugs of stainless steel or Teflon are used in the ends of the columns to retain the packing material. The plugs must be homogeneous to ensure the uniform flow of solvent through the column. It is important in some separations involving liquid partition and ion-exchange that the column temperature us thermostatically controlled during the analysis.

Column Packing Procedure

Columns may be purchased already packed from commercial companies with specified packing material structure and dimensions. Many workers, however, prefer to pack their own columns since this is cheaper than purchasing pre-packed columns. Several methods are available for packing columns and the method used will depend on the nature of the packing material and the dimensions of the particles. The major priority in the packing of a column is to obtain a uniform bed of material with no cracks or channels. Rigid solids and hard gels should be packed as densely as possible, but without fracturing the particles the packing process. The most widely used technique for column packing is the high-pressure slurring technique. A suspension of the packing is made in a solvent of equal density to the packing material. The slurry is then rapidly pumped at high pressure into a column with a porous plug at its outlet. The resulting bed of packed material within the column can then be prepared for use by running the developing solvent through the column, hence equilibrating the packing with the developing solvent. When hard gels are packed, it is necessary for them to be allowed to swell first in the solvent to be used in the chromatographic process before packing under pressure. Soft gels cannot be packed under pressure and have to be allowed to pack from slurry in the column under gravitational sedimentation only, in a similar way to the packing of columns for conventional column chromatography.

Chromatographic Solvent (Mobile Phase)

The choice of mobile phase to be used in any separation will depend on the type separation to be achieved. Isocratic separations may be more made with a single solvent, or two or more solvents mixed in fixed proportions. Alternatively, a gradient elution system may be used where the composition of the developing solvent is continuously changed by use of suitable gradient elution system may be used where the composition of the developing solvent is continuously changed by use of a suitable gradient programmer. In the majority of cases this involves the use of two pumps. All solvents for use in HPLC systems must be specially purified since traces of impurities can affect the column and interfere with the detection system. This is particularly the case if the detection system is measuring absorbance at below 200 nm. Purified solvents for use in HPLC systems are available commercially, but even with these solvents a 1 to 5 μm micro filter is generally introduced into the system prior to the pump. It is also essential that all solvents are degassed before use otherwise gassing tends to occur in most pumps. It tends to be particularly bad for aqueous methanol and ethanol solvents, gassing (the presence of air bubbles in the solvent) can alter column resolution and interface with the continuous monitoring of the column effluent. Degassing may be carried out in several ways; by warming the solvent, by stirring it vigorously with a magnetic stirrer, subjecting it to a vacuum, ultrasonic vibration, or by bubbling helium gas through the solvent reservoir.

Pumping Systems

The pumping system is one of the most important features of an HPLC system. There is a high resistance to solvent flow due to the narrow columns packed with small particles, and high pressures are therefore required to achieve satisfactory flow rates. The main feature of a good pumping system is that it is capable of outputs of at least 3.4 × 1.07 Pa (5000 psi) and ideally there must be no pulses of flow through the system. There must be a flow delivery of at least 10 cm^3 min^{-1} for normal analysis, and up to 30 cm^3 min^{-1} for preparative analysis. All materials in the pump should be chemically resistant to all solvents. Various pumping systems are available which operate on the principle of constant pressure or constant displacement.

Constant pressure pumps produce a pulse less flow through the column, but any decrease in the permeability of the column will result in lower flow rates for which the pump will not compensate. These pumps will not compensate. These pumps operate by the introduction of high-pressure gas into the pump, and the gas into the pump, and the gas in turn forces the solvent from the pump chamber into the column. The use of an intermediate solvent between the gas and the eluting solvent reduces the chances of dissolved gas directly entering the eluting solvent and causing problems during the analysis.

Constant displacement pumps maintain a constant flow rate through the column irrespective of changing conditions within the column. One form of constant displacement pump is a motor-driven syringe type pump where a fixed volume of solvent is forced from the pump to the column by a piston driven by a motor. Such pumps, as well as providing uniform solvent flow rates, also yield a pulse less solvent flow which is important as certain detectors are sensitive to changes in solvent flow of constant displacement pump. The piston is moved by a motorized creak and entry of solvent from the reservoir to the pump chamber and exit of solvent to the narrow band onto the column. There are two methods which are generally used. The first method makes use of a micro syringe designed to withstand high pressures. The sample is injected through a septum in an injection port, either directly onto the column packing or onto a small plug of inert material immediately above the column packing. This can be done the pump may be turned off before injection, and when the pressure has dropped to near atmospheric, the injection is made and the pump switched on again. This is termed a stop flow injection. The second method of sample introduction is by use of a loop injector. This consists of a metal loop of small volume which can be filled with the sample. By means of an appropriate value, the eluent from the pump is channeled through the loop, the outlet of which leads directly onto the column. The sample is column is regulated by check values. On the compression stoke solvent forced from the pump chamber into the column. During the return stroke the exit check value closes and solvent is drawn in via. The entry value to the pump chamber, ready to be pumped onto the column on the next compression stoke. Such pumps produce pulses of flow and pulse dampness are usually incorporated into the system to minimize this pulsing affect. All constant displacement pumps have in-built safety cut-out mechanisms so that if the pressure within the chromatographic systems changes from pre-set limits the pump is inactivated.

Practical Procedure

The correct application sample onto a HPLC column is another particularly important factor in achieving successful separations. Ideally the sample ought to be introduced as an infinitely thus flushed onto the column by the eluent, without interruption solvent flow to the column. Automatic versions of loop injectors are commercially available.

Repeated application of highly impure samples such as sera, urine, plasma, or whole blood, which have preferably been deproteinized, may eventually cause the column to lose its resolving power. To prevent this occurrence a guard column is installed between the injector and the analytical column. This guard column is a short (2–10 cm) column. Of the same internal diameter, and packed with similar material to that present in the analytical column. The packing of the guard column can be replaced at regular intervals.

Applications

The wide applicability, speed, and sensitivity of HPLC have resulted in it becoming the most popular form of chromatography and virtually all types of biological molecules have been purified using the approach. Reverse phase partition HPLC is particularly useful for the separation of polar compounds such as drugs and their metabolites, peptides, vitamins, polyphenols, and steroids. Prior to the advent of this form of chromatography, the separation of such polar compounds was not easily accomplished and often required pre-derivatization to less polar compounds.

HPLC has probably had the biggest impact on the separation of oligopeptides and proteins. Instruments dedicated to the separation of proteins have given rise to the technique of fast protein liquid chromatography (FPLC). There are no unique principles associated with FPLC; it is simply based on reverse phase and ion-exchange chromatography and on chromato-focusing. Micropore glass-lined stainless steel columns 1 mm diameter and 2.5 cm long have recently been developed which enable very small amounts of sample to be used with separation taking as little as 10 min.

Ion Exchange Chromatography

Introduction

Electrostatic attraction of oppositely charged ion on a polyelectrolyte surface forms the basis of ion exchange chromatography. Typical systems include the synthetic resin polymers, such as the strongly acidic cation exchanger Dowex-50, a polystyrene sulfonic acid, and a strongly basic anion exchanger Dowex-1, a polystyrene quaternary ammonium salt. Cellulose derivates such as carboxymethyl cellulose (CMC) and diethyl amino ethyl cellulose (DEAE) exchangers have been very successfully used in protein purification.

The basic principle involves an electrostatic interaction with the exchanging ion and the normal charge on the surface of the resin. These reactions are considered to be equilibrium-processed and involve diffusion of a given ion to the resin surface and then to the exchange site, the actual exchange, and finally diffusion away from the resin. The rate of movement of a given ionizable compound down the column is a function of its degree of ionization, the concentration of other ions, and the relative affinities of the various ions present in the solution for charged sites on the resin. By adjusting the pH of the eluting solvent and the ionic strength, the electro statically held ions are eluted differently to yield the desired separation.

The separation in ion exchange chromatography is obtained by reversible adsorption. Most ion exchange experiments are performed in two main stages.

The first stage is sample application and adsorption. Unbound substances can be washed out from the exchanger bed using a column volume of starting buffer. In the second stage, substances are eluted from the column, separated from each other. The separation is obtained since different substances have different affinities for the ion-exchanger according to the differences in their charge. These affinities can be controlled by varying conditions such as ionic strength and pH. Ion- exchange chromatography is capable of separating species with very minor differences in properties, such as two proteins differing by only one amino acid—it is most widely used technique for separating, identifying, and quantitating the amounts of each amino acid in a mixture. The entire procedure has been automated, so that the elution, collection of fractions, analysis of each fraction, and recording of data are performed automatically in an amino acid analyzer.

In ion exchange chromatography one can choose whether to, bind the substances of interest, or to adsorb out contaminants and allow the substance of interest, to pass through the column. Generally, it is more useful to adsorb the substance of interest, since this allows a greater degree of fractionation.

Ion exchange may be carried out in a column or by a batch procedure. Both methods are performed in definite stages. These stages are: equilibration of the ion exchanger, addition and binding of sample substances, change of conditions to produce selective desorption, and regeneration of the ion exchanger.

Theory

The matrix many commercial ion-exchangers which have been successfully employed for separation of biological materials are made by co-polymerizing styrene with divinylbenzene, however, cross-linking of molecules occurs and this produces an insoluble resin. Various degrees of cross linkage may be obtained by co-polymerizing varying proportions of divinylbenzeneand styrene. The higher the amount of divinylbenzene, employed with respect to the styrene, the greater the degree of cross- linkage obtained. Resins with a low degree of cross-linking are more permeable to high molecular weight compounds than are highly cross-linked ones, but they are also less rigid and swell more when paced in a buffer. These swelling characteristics must be taken into account when a column is prepared. Sulphonation of cross- linked polystyrene resin such as Dowex 50 which is strong acidic exchangers. The SO_3 H group is ionized at all except very low pH values. An analogous basic exchanger may be prepared by reacting cross-linked polystyrene with chloromethyl ether, and then reacting the chloro groups with tertiary amines. These $-CH_2N+ (CH_3)_3Cl^-$ groups are ionized at all but very alkaline pH values.

Ion-exchange chromatography
(anion exchange)

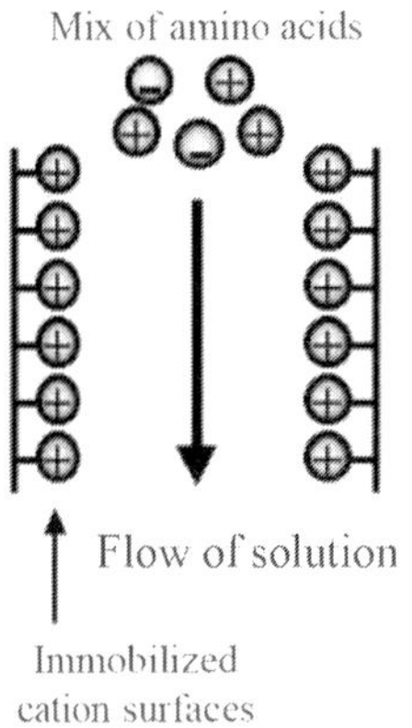

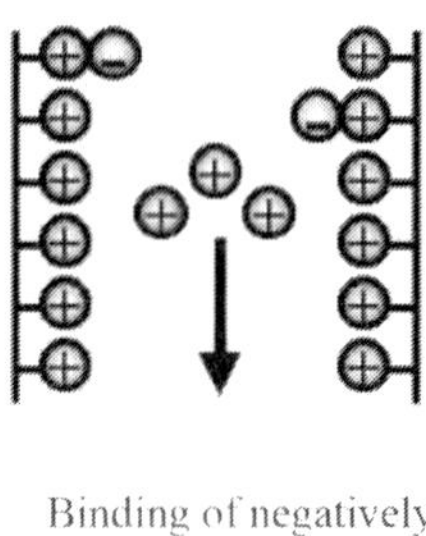

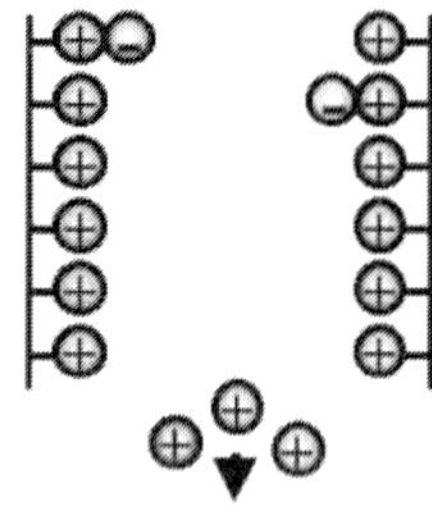

Chemically modified celluloses have proved to be a particularly useful alternative to the polystyrene-based exchangers. Cellulose is a high molecular weight compound which can be obtained in a highly pure state. Carboxymethyl cellulose (CM–CH_2 OCH_2 COOH cellulose), and DEAE-cellulose (–$CH_2OCH_2CH_2N(CH_2CH_3)_2$) are examples of the main derivatives of practical value.

Ionizable Groups

Ionizable groups are charged groups that are attached to the matrix and the type of group defines the nature and strength of the ion exchanger. These groups may be either anionic or cationic, according to the nature of their affinity for either negative or positive ions. For example, the cation exchanger materials exchange positive ions, so it is the charge carried by the exchangeable ion which decides whether a material is anionic or cationic and not the charge carried on the matrix. These two types can be further divided into materials that contain strongly ionized groups, such as –SO_3H and –NR_3, and the weakly ionized groups, such as –COOH, –OH and –NH_2. The strong ion exchange resins are completely ionized and exist in the charged form except at extreme pH values:

$$-SO3H \longrightarrow SO^-_3 + H^+$$

$$-NR3OH \xrightarrow{+} NR_3 + OH^-$$

The weak ion exchange materials, on the other hand, contain groups whose ionization is dependent on the pH, and they can only be used at maximum capacity over a narrow pH range.

$$-COOH \rightleftarrows -COO^{-} + H^{+}$$

$$-NH_3^{+} \rightleftarrows -NH_2 + H^{+}$$

As a rough guide, resins containing carboxyl groups have a maximum capacity above pH 6, while those with amino groups are effective below pH6.

The number of ionizable groups determine the capacity of the ion exchanger. The total capacity is the number of ionizable groups per gram of material, whereas the available capacity is the amount of a given molecule that can bind under defined experimental conditions. In the case of some materials, large molecules may be unable to penetrate the matrix and can only react with charged groups on the surface. In this case, the available capacity will be considerably less than the total capacity.

Although ion exchange materials are claimed to be monofunctional, in that only the ion exchange process is used in separation, in practice some molecular sieving and adsorption can occur. The adsorption is small, but it can sometimes be used to separate closely related compounds.

Elution of Bound Ions

The bound ions can be removed by changing the pH of the buffer. For example, as the pH of a protein moves towards its isoelectric point, the net charge decreases and the macro-molecule is no longer bound. Separation is achieved as other charged proteins remain on the column alternatively, ions can be removed by increasing the ionic strength, when high concentrations of ions in the solvent displace the bound ions by increasing the competition for the charged groups of the ion exchange material. The pH or ionic strength can be altered sharply, by changing the eluting buffer, or, gradually, by means of a gradient.

Preparation of Material

Ion exchange materials are first allowed to swell in the buffer. The ion exchange material is then obtained in the required ionic form by washing with the appropriate solution. For example, the H^{+} form of a cation exchange resin is obtained by washing the material with hydrochloric acid then water until the washing are neutral. Similarly, the Na^{+} form is prepared by washing the resin with sodium chloride or sodium hydroxide then water as above.

The final stage before preparing the column is to equilibrate the material by stirring with the eluting buffer.

The Separation of Amino Acids by Ion Exchange Chromatography

Materials

1. Chromatography column (20 cm × 1.5 cm) 5
2. Strongly acidic resin 30 g.
3. Hydrochloric acid (4 mL/L) 1 L.
4. Hydrochloric acid (0.1 mL/litre) 4 L.
5. Glass wool.
6. Amino acid mixture 10 mg.

   (dissolve aspartic acid, histidine of each and lysine in 0.1 mol/L HCl to a final concentration of 2 mg/mL)
7. Tris–HCl buffer (0.2 mol/L, pH 8.5) 3 L.
8. Sodium hydroxide (0.1 mol/L) 2 L.
9. Separating funnels (500 mL) 10.
10. Acetate buffer (4 mol/L, pH 5.5) 250 mL.
11. Ninhydrin reagent (dissolve 20 g of ninhydrin and 3 g of 1 Lhydrindantin in 750 mL of methyl cello solve and add 250 mL of acetate buffer). Prepare fresh and store in a brown bottle.
12. Methyl cello solve (ethylene glycol monomethyl ether) 1 L.
13. Ethanol (50 %v/v) 1 L.
14. Ninhyrin (2 g/Lin acetone). (Care: carcinogenic!) 100 mL.
15. Oven at 105 °C 1.

Method

Preparation of the Column

Gently stirs the resin with 4 mol/L HCl until fully swollen (15–30 mL/g dry resin). Allow the resin to settle, and then decant the acid. Repeat the washing with 0.1 mol/L HCl, resuspend in this solution, and prepare the column as previously described.

Elutions of Amino Acid

Carefully apply 0.2 mL of the amino acid mixture to the top of the column, open the tap and allow the sample to flow into the resin. Add 0.2 mL of 0.1 mol/L HCl, allow to flow into the column as before, and repeat the process twice. Finally, apply 2 mL of 0.1 mol/ litre HCl to the top of the resin and connect the column to a reservoir containing 500 mL of 0.1 moL/LHCl. Adjust the height of the reservoir to give a flow rate of about 1 mL/min and collect a total of forty 2 mL fractions. Test five of the tubes at a time for the presence of amino acids by spotting a sample from each tube on to a filter paper: dip this in the acetone solution of ninhydrin and heat in an oven at 105 °C. If amino acids are present, they will show up as blue spots on the filter paper. When the first amino acid has been eluted, remove the reservoir of 0.1 mol/L HCl and allow the level of acid to fall to just above the resin. Run 2 mL of 0.2 mol/L tris–HCl buffer (pH 8.5) on to the column, connect to a reservoir of this buffer and continue with the elution until the third amino acid is removed from the column.

Detection of Amino Acids

Adjust the pH of each tube to 5 by the addition of a few drops of acid or alkali. Add 2 mL of the buffered ninhydrin reagent and heat in a boiling water bath for 15 min. Cool the tubes to room temperature, add 3 mL of 50% v/v ethanol and read the extinction at 5570 nm after allowing the tubes to stand for 10 min, set up the appropriate blanks and standards, and plot the amount of amino acid in each fraction against the volume eluted.

Applications

Ion exchange has proved to be one the major methods of fractionation of labile biological substances. It is used most widely for protein and peptide separations, enzyme purification, hormone preparation, studies, carbohydrate and lipids separations and vitamin and co-enzyme work etc. the special characteristic of ion exchange is that the separation is based on charge, so that when it is combined with techniques which use other criteria for separation, such as gel filtration (size) or affinity chromatography (bio selectivity), it is a very powerful analytical and preparative tool.

Gel Filtration

The technique of separating molecules of different size by passage through a gel column is called gel filtration. Small molecules can enter the gel but larger molecules are excluded from the cross- linked network. This means that the accessible volume of solvent is very much less for molecules totally excluded from the gel than for small molecules which are free to penetrate. The separation of molecules by gel filtration is illustrated in the diagram. First the mixture of large and small molecules is placed on top of the column. As they pass down the column, the small molecules diffuse into the gel ad follow a longer path than the large molecules, which are completely excluded from the gel particles.Eventually, complete separation occurs, with the large molecules leaving the column first and the smaller ones last.

A gel filtration separation can be performed in the presence of essential ions or cofactors, detergents, urea, at high or low ionic strength, at 37 °C or in the cold room according to the requirements of the experiment.

Principle

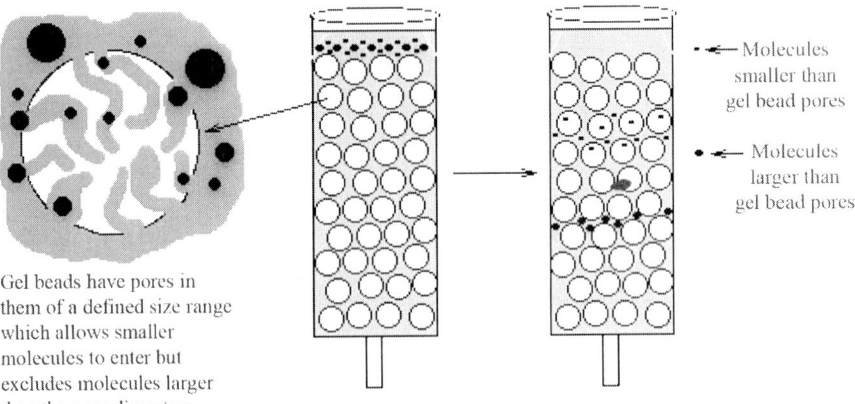

The technique of separating molecules of different size by passage through a gel column is called gel filtration. Sephadex is the most popular of the gel materials. A polysaccharide, dextran, is carefully cross-linked to give small beads of a hydrophilic, insoluble nature which when placed in water swells considerably to form an insoluble gel. Sephadex is the commercial name for the gel prepared thus.

Sephadex is prepared from polysaccharide dextran which has been synthesized by the action of a bacterium on sucrose. Each glucose residue contains 3 hydroxyl groups giving the dextran a polar character. The cross-linking reaction is accompanied with epichlorohydrin.

$CH_2\,CH\,CH_2Cl$

\O/

Epoxy group

By adjusting the conditions, the amount of cross- linking, and thus the size of pores, can be carefully controlled. Sephadex gels are insoluble in all solvents and are stable in bases, weak acids, water and mild reducing and oxidizing agents. Long exposure to 0.1N HCl or strong oxidizing agents will cause breakdown of the gel granules. Temperature of about 120 °C should be avoided, too. In addition to water in which they are commonly soluble, sephadex gels swell in glycel, dimethyl sulfoxide,and form amide but not in glacial acetic acids, methanol, or ethanol.

These gels are classified by the amount of "water regained" which is a function of the looseness of the structure or the extent of cross linking. By sephedex gel, the useful fractionation range of molecules is only approximate since separation depends upon the shape, charge to a minor extent and size of the molecules. Sephadex G-25 and G-50 are made in several particle sizes and we find coarse, medium, and fine grades. Fine grades are suitable for most laboratory work requiring high resolution, while the coarse material is convenient for preparative work with large columns since this gives a higher flow rate.

Various other materials which give improved resolution have been developed e.g. sepharose. Sepharose is stable from pH 4.0–10.0 and can be used over the temperature range 0–30 °C. Three different types of sepharose are available which have different concentrations of agarose. Sepharose CL is stable over pH 3.0 to 14.0and can be used upto temperatures of upto 70 °C. These are some porens gels and are used to fractionate very large molecules such as nucleic acids and viruses. After use, the columns can be washed in water and stored in the cold room for quite a time provided traces of preservative such as phenol or chloroform is added to prevent bacterial growth. Another material used for column chromatography is "sephacryl." It is a polymer of allyldextran covalently crossed lambda linked with N–N′ methylene-bisacrylamide and can be obtained in a preswollen form. The material gives much faster flow rates than sephadex or biogel and 5200 can be used to separate proteins with a molecular weight between 5000 and 2,50,000. The molecular weight of proteins can be determined from elution volume after calibration of column with known molecular weight markers.

Another gel chromatography material that is used for molecular extrusion chromatography is Biogel-P of different types. They are made from polyacrylamide.

Principle of Separation

Theory of Gel Filtration

The total volume of the column (V_T) is the sum of volume of the gel matrix (V_g), the volume of water inside the gel particle (V_i) and the volume of the water outside the gel grains (V_o).

i.e. $V_T = V_o + V_g + V_i$,

V_o i.e. void volume is the volume of the liquid required to elute compounds that are completely excluded from the gel grains, and is known. V_i can be calculated from the knowledge of dry weight of the gel (a) and the water regained (W_r) e.g. $V_i = a \times W_r$. The elution volume V_e of a compound is the volume required to elute that compound from a column i.e. $V_e = V_i + K_{d}, V_i$ where K_d indicates the fraction of the inner volume accessible to a particular compound and is independent of the geometry of the column. i.e. $K_d = V_e - V_o$. Substituting the value of V_i in the above equation we get,

$V_e - V_o$

$K_d = aW_r$

If $K_d = 0$, then $V_e = V_o$ i.e. the elution volume would be the void volume.

If $K_d = 1$

$V_e - V_o = V_i$,

Or $V_e = V_i + V_0$

If a molecule is completely eluded from the gel, then $K_d = 0$ and $V_e = V_0$, while if a molecule has complete accessibility to the gel, then $K_d = 1$.

Molecules therefore have a K_d value between 0 and 1. If $K_d > 1$, then adsorption of the compound on the gel occurs.

By plotting a graph of elution volume against molecular weight taking different markers and comparing the elution volume of the test compound with the standard graph, its corresponding molecular weight can be known.

Applications

Gel chromatography has been widely used by biochemists to separate proteins, peptides, nucleic acids, polysaccharides, enzymes, and hormones alike and by polymer chemists to characterize molecular weight distribution in polymeric mixtures. The other applications of gel filtration chromatography are as follows:

Desalting

One of the common separations of salts and small molecules from macromolecules. The large difference in distance coefficient for this separation makes it possible to use simple columns with high flow rates.e.g. With sephadex G-25, solute molecules having molecular weights over 5000 are eluted in a volume V_o, smaller molecules with molecular weights below 1000 will be eluted following the macromolecules inversely in order of their size. One of the advantages of this method of desalting is that the macromolecules are eluted with essentially no dilution.

Concentrating

The diluted solution of macromolecules with molecular weights higher than the exclusion limit may be readily concentrated by utilizing the hygroscopic nature of dry gels.

Sephadex G 200 absorbs 20 times its weight of water, although G-25 is preferred because of its more rapid action. Since salt molecules are imbibed to the same extent as water, their concentration in supernatant solvent remains essentially unchanged and ionic strength and pH of macromolecular solution remains essentially unaltered. This is an important advantage in separation of proteins which are easily denatured by change in temperature.

Fractionation

Components separation of mixture of closely related materials having small difference in *K*-values will require long columns, slow flow rates and long times. The difference of molecular weights in a complex mixture is often sufficient information to check the solution or to give a rough idea of its composition. It has been observed and can be predicated that elution volume is related to the molecular weight in a simple fashion. $V_e = (A - B)$ log molecular weight. Where A and B are constants, whose values are determined from a plot of V versus log molecular weight for several known compounds. The equation is valid over a considerable range of molecular weights, provided all compounds are closely related. Values for A and B must be redetermined for each column in each set of specific conditions and for different classes of compounds.

Separation of Proteins by Gel Filtration

S-200 can be used to separate proteins with a molecular weight between 5000 and 2, 50,000. The molecular weight of a protein can be determined from the elution volume after calibrating the column with known molecular weight markers.

Materials

1. Sephacryl S-200 columns (20 cm × 2 cm) equilibrated with the elution buffer-5.
2. Elution buffer (0.5 mol/L NaCl in 0.1 mol/litre tris-HCl buffer, 3 L pH 8.0).
3. Buffer reservoirs 5.
4. Fraction collectors 5.
5. Molecular weight standards (5 mg of each protein per ml of elution buffer).

Standard	mol.wt.	Log mol.wt.
Cytochrome C	12,400	4.09
Ovalbumin bovine serum	45,000	4.65
Y-Globulin	210,000	5.32
Blue dextran	2,000,000	-

6. Lactate dehydrogenises 1 mL.
7. Spectrophotometer 5.
8. Reagents for the assay of glucose.
9. Reagents for the assay of lactate dehydrogenises.

Method

Running the Column

Open the tap on the column and allow the level of the elution buffer to fall until it is just above the top of the gel. Mix 0.1 mL of lactate dehydrogenises solution and 0.4 mL of the markers and place this on top of the column. Allow the mixture to enter the gel, add a small volume of the elution buffer until the markers are clear of the top then connect the buffer reservoir. Adjust the height of the reservoir to give a flow of about 50 ml/h and collect 3 mL fractions using the fraction collector. Locate the position of the blue dextran and the glucose and examine the fractions between these compounds for the presence of the proteins.

Detection of Compounds

Blue dextran: extinction at 625 nm Y-globulin and ovalbumin: extinction at 280nm cytochrome c: extinction at 412 nm glucose and lactate dehydrogenase: as per the standard method prescribed.

Molecular Weight Determination

Calibrate the column by determining the volume at which the blue dextran is eluted (V_0) and the volume when the glucose appears ($V_0 + V_i$). Determine the elution volume (V_e) of the three marker proteins and the LDH and calculate the K_d values. Plot a graph of K_d against log10 mol.wt. For each marker and calculate the apparent molecular weight of the LDH.

Spectrophotometry

Demonstration of Spectrophotometer

Spectrophotometer consists of a light source, a monochromator (device to produce a monochromatic beam of light from any point on the spectrum), and a vessel containing the solution and positioning it in the light beam, and a detector with its amplifier. The light source used for the visible spectrum is the tungsten lamp, usually a single filament car headlamp bulb. This gives a peak energy output at about 950nm and coves the range 320–3000 nm. The electric supply must be constant as the radiant energy must not vary. This is provided by an accumulator or car battery. A constant voltage transformer or stabilized power pack is used in most instruments. The source of UV light is the hydrogen or deuterium discharge tube which supplies light of 180–350 nm. A stabilized power supply is essential for these lamps. The tube has a quartz window and when adjustments are made this should not be touched. If a more intense source is required the mercury arc or xenon lamp is used.

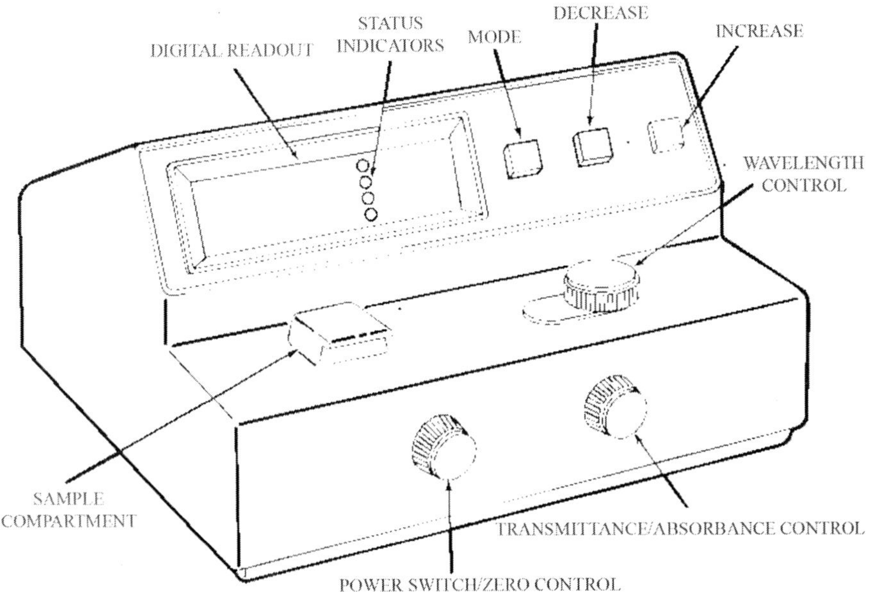

Infra-rod radiation is supplied by a Nernst filament, rod of oxides of zirconium and thorium, which requires a heater winding for starting, the operating temperature being about 1350 °Ca stabilized voltage is required.

The visible or UV type employs either a glass or quartz prism or a diffraction grating in the monochromator. Some instruments are both prism and grating to improve resolving power. When working with the UV region, silica cells must be used as glass is not transparent to UV light.

Operation

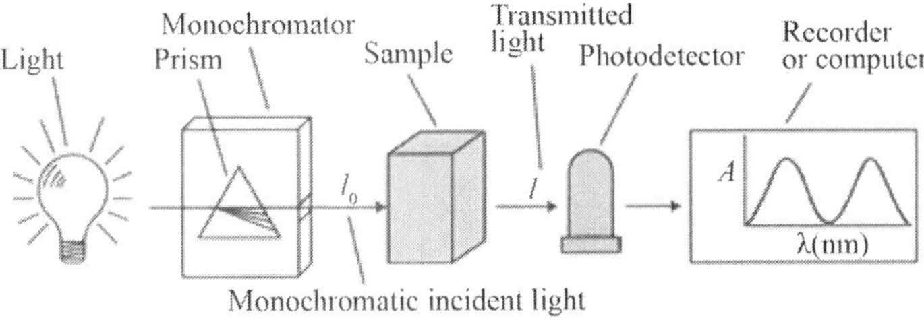

The instrument is turned on and allowed to warm up for a few minutes. The illumination must be stable before starting. Most instruments have "dark current" controls which allow the balancing of the electronics with no light entering the photo cell. This is done after a wavelength has been selected. With a reagent blank in the light path the galvanometer or potentiometer is set at zero. The blank is then replaced by the test solution and the absorbance or transmission is noted. The instrument should be zeroed with the blank in position with each successive wavelength.

Types of Spectroscopy

1. Visible spectrophotometry.
2. Ultraviolet spectrophotometry.
3. Infra-red spectrophotometry.
4. Circular dichroism spectrophotometry.
5. Spectrofluorimetry.
6. Luminometry.
7. Atomic/flame spectrophotometry.
8. Electron spin resonance spectrophotometry.
9. Nuclear magnetic resonance spectrophotometry.
10. Mass spectrometry etc.

Principle of Spectrophotometer

A spectrophotometer is employed to measure the amount of light that a sample absorbs. The instrument operates by passing a beam of light through a sample and measuring the intensity of light reaching a detector.

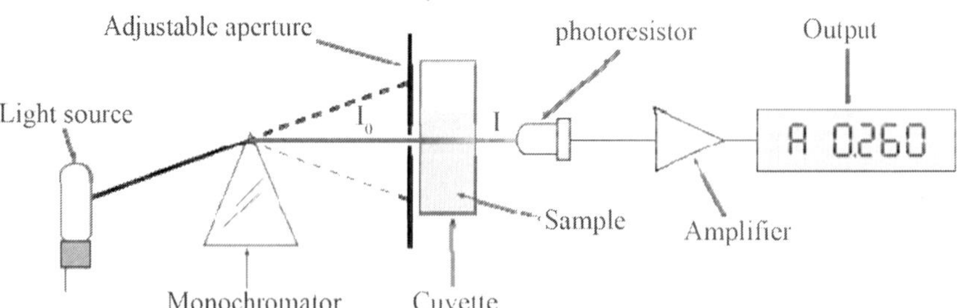

A spectrophotometer consists of two instruments, namely a *spectrometer* for producing light of any selected color (wavelength), and a *photometer* for measuring the intensity of light. The instruments are arranged so that liquid in a cuvette can be placed between the spectrometer beam and the photometer.

The amount of light passing through the tube is measured by the photometer. The photometer delivers a voltage signal to a display device, normally a galvanometer. The signal changes as the amount of light absorbed by the liquid changes.

If development of color is linked to the concentration of a substance in solution, then that concentration can be measured by determining the extent of absorption of light at the appropriate wavelength.

When monochromatic light (light of a specific wavelength) passes through a solution there is usually a quantitative relationship (Beer's law) between the solute concentration and the intensity of the transmitted light, that is,

$I = I_0 \times 10^{-kcl}$,

where I_0 is the intensity of transmitted light using the pure solvent, $I$is the intensity of the transmitted light when the colored compound is added, c is concentration of the colored compound, l is the distance the light passes through the solution, and k is a constant. If the light path l is a constant, as is the case with a spectrophotometer, Beer's law may be written,

$I/I_0 = 10^{-kc} = T$,

where k is a new constant and T is the transmittance of the solution. There is a logarithmic relationship between transmittance and the concentration of the colored compound. Thus,

Log T = log $1/T = K_c$ = optical density (O.D.).

The O.D. is directly proportional to the concentration of the colored compound. Most spectrophotometers have a scale that reads both in O.D. (absorbance) units, which is a logarithmic scale, and in percentage transmittance, which is an arithmetic scale. As suggested by the above relationships, the absorbance scale is the most useful for colorimetric assays.

Experiment

Verification of Lambert Beer's Law

Materials and Reagents

1. Spectrophotometer.
2. A pair of Quartz cuvettes (3 or 1 mL capacity).
3. Test tubes.
4. Tissue paper/filter paper.
5. 0.5 mg/mL solution of BSA.

Procedure

i. Make different dilutions of BSA solution in different test tubes ranging from 0.05 to 0.1 mg/mL.
ii. Measure the absorbance of above solutions at the above measured λ_{max}.
iii. Draw a graph between concentration along x axis and absorbance along *y* axis.
iv. Check the linearity of the curve.

Observations

The range of concentration in which the graph is linear, the Lambert Beer's law is obeyed.

High Speed Centrifuge and Ultra-Centrifuge

Introduction

The physical techniques most responsible for current understanding of cellular makeup and operation are those involving the centrifuge. Wide varieties of these instruments are available ranging in capacity from those handling 0.2 mL and less to those accommodating thousands of liters with relative case. Some are crudely controlled with regard to speed and temperature whereas in other these parameters are regulated within limits of less than 5%.

In its simplest form a centrifuge is composed of a metal rotor with holes in it to accommodate a vessel of liquid and a motor or other means of spinning the rotor at a selected speed. All the other parts found in today's modern centrifuge are merely accessories used to perform various useful tasks and maintain the environment within which the rotor operates.

Principle

Centrifugation separation techniques are based upon the behavior of particles in an applied centrifugal field. Particles which differ in density, size, or shape sediment at different rates in a centrifugal field. The rate of sedimentation of a spherical particle is dependent upon the applied centrifugal field, density, radius of the particle and the viscosity of the suspending medium. These relationships are expressed by Stoke's law.

$$U = 2/9\, r^2 p \frac{(lp - lm)}{n} \times g$$

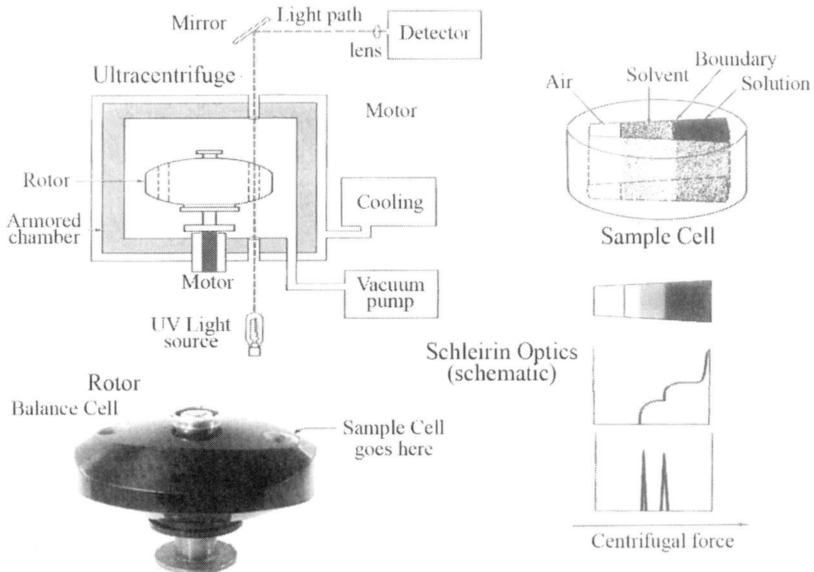

Where

u = sedimentation rate or velocity of sphere,

ℓp = density of the particle,

ℓm = density of the suspending medium,

$r\ p$ = radius of the particle,

n = viscosity of the suspending medium

g = gravitational field

2/9 = shape factor constant for a sphere.

If the density of medium and density of particle are equal, then the sedimentation rate will be zero. But this sedimentation rate will also depend upon its sedimentation rate. Hence, even if the particles have similar density, slight differences in their size can make large differences in their sedimentation velocity.

Classification

Centrifugation techniques can be mainly classified into two types namely,

a. **Preparative centrifugation and**

b. **Analytical centrifugation**

Preparative Centrifugation

This is concerned with the actual isolation of biological material for subsequent biochemical investigations. Very large amounts of material may be involved when harvesting, for example, microlical cells from batch or continuous culture, plant and animal cells from tissue culture and plasma from blood. Relatively large amounts of cellular particles may also be isolated in order to study their morphology, composition and biological activity. It is also possible to isolate such biological macromolecules as DNA and proteins, from preparations which have received some preliminary purification, for example fractional precipitation.

Analytical Centrifugation

It is used for the study of pure or virtually pure macromolecules requires for e.g.: ribosome. It requires very small amount of material.

Preparative centrifugations are the more commonly used ones. This can be further classified into four groups:

i. Differential centrifugation.
ii. Rate zonal centrifugation.
iii. Equal density or isopycnic.
iv. Equilibrium isodensity centrifugation.

i. Differential centrifugation

This is based on the difference on sedimentation rate of particles of different size and density. The material to be separated is centrifugally divided into a number of fractions by the step wise increase of applied centrifugal field. The centrifugal field at each stage is chosen so that a particular type of material sediments, during the pre-determined time of centrifugation to give a pellet. At the end of each stage the pellet is washed to obtain a pure fraction. The sedimentation of particles during centrifugation varies depending upon their size and shape.

It is the most commonly used method for the isolation of cell organells from homogenized tissue. The main demerit is that the liquid density gradient has to be performed when it is required and this may be a time-consuming process.

ii. Rate-zonal centrifugation

This is also called as s-zonal centrifugation. This involves carefully layering a sample solution containing particles to be separated, on top of a performed liquid density gradient whose density continuously increases towards the bottom of the centrifuge tube. The sample is prevented from pre-mature sedimentation by the steep positive density gradient beneath it. The sample is then centrifuged until the desired degree of separation is affected i.e. complete sedimentation should not be there, but the particles should form discrete zones or bands consisting of particles characterized by their sedimentation rate. Here, to achieve separation density of particles must be higher than the density gradient.

This method is used for the separation of RNA–DNA hybrids, ribosomal subunits, and other cell components.

iii. Equal density (isopycnic) centrifugation

Here density gradient may or may not be used. In the absence of a continuous density gradient, the sample is initially centrifuged to sediment particles heavier than the ones required. After separation and rejection of heavier particles the sample is suspended in a medium having a density equal to the fraction to be isolated. After further centrifugation the lighter and heavier particles than the required ones will be at the meniscus and base of the tube, respectively, while the required one will be in an intermediate position. In the presence of density gradient it will be same as that of rate zonal method. But the main difference between the two methods is that for equal density centrifugation, the centrifugation is continued until the desired particles have reached their equal density or isopycnic position in the gradient. This is not required in rate separation.

Equal density or isopycnic position: means where the buoyant density of the particle and the density of the gradient are equal. At this point no further sedimentation occurs.

iv. Equilibrium isodensity centrifugation

In this method, for the preparation of gradient, salt of heavy metals like Co or rubidium or sucrose is used. The sample is mixed homogeneously. The sample is mixed with a concentrated solution of cell and centrifuged. This gives a concentration gradient of Cs c l and hence a density gradient due to large mass of Cs ion. Redistribution of particles then takes place in this gradient.

This is the most commonly used analytical centrifugation method. It is also used to separate and analyze human plasma lipoproteins.

Types of Preparative Centrifuge

Preparative centrifuge can be classified into three major groups:

1.General purpose centrifuge

2.High speed centrifuge, and

3.Ultra centrifuge

General Purpose Centrifuge

General purpose centrifuge has a maximum speed of 6000 rpm. Used for collecting substances that sediment rapidly, like yeast cells, RBC's etc.

High Speed Centrifuge

These can operate up to speeds of 20,000 to 25,000 rpm. These are generally equipped with refrigeration equipment to cool the rotor chamber. Two types are there. The first is relatively simple, high capacity, continuous flow centrifuge. If consists of a very fast electric motor, which is connected to a long tubular rotor. This is used for the harvesting of yeast or bacteria from large cultures.

The culture is siphoned or pumped into the bottom of the spinning rotor. As the cultures move up towards the top of the rotor, microorganisms are sedimented against the rotor walls and the clarified culture medium exits through an effluent part. After all the medium has been passed through the centrifuge the rotor is disassembled and the compared cells are removed with long spactula.

The second type consists of lower capacity refrigerated equipment. A broking device is also used here to decrease the time required for rotor declaration at the end of centrifugation, it is used to collect microorganisms, cellular debris, cells, large cell organelles, immunoprecipitates, etc.

The main demerit is that it cannot sediment viruses, small organells like ribosome's.

Ultra-Centrifuge

It can operate at a speed of 75,000 rpm. It has four major parts.

a. drive and speed control
b. temperature control
c. vacuum system, and
d. rotors.

Uses

1. It permits fractionation of sub cellular organells which can be seen only with the help of electron micrographs.
2. It can be used for the isolation of viruses.
3. DNA, RNA and protein can be carefully analyzed.

Flame Photometry

Flame photometry is an atomic emission technique which is used for the detection of metal salts mainly Sodium (Na),Potassium(K), Lithium(i), Calcium (Ca), and Barium(Ba). It is a simple and inexpensive method.

Principle

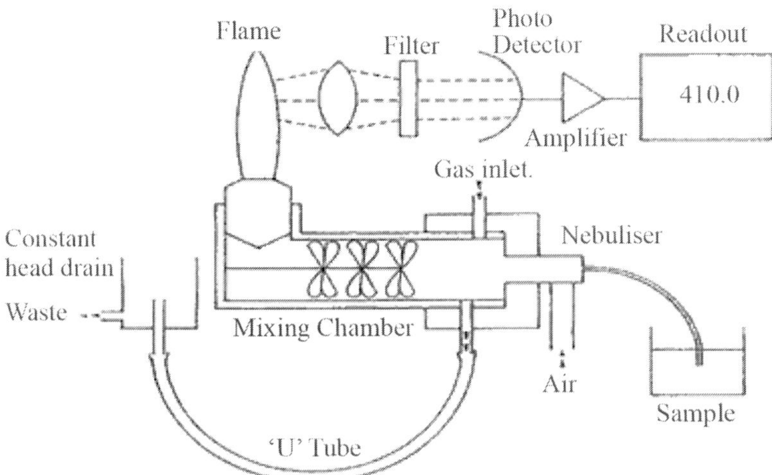

Flame photometry quantitatively analyzes the metals by measuring the flame emission of solutions containing the metal salts. Solutions are aspirated to the flame. The hot flame evaporates the solvent, atomizes the metal and excites a valence electron to an upper state. Light is emitted at characteristic wave lengths for each metal as the electrons return to the ground state. Optical filters are used to select the emission wavelength monitored for the analyte species. Comparison of emission intensities of unknowns to either that of standard solutions, or to those of an internal standard, allows quantitative analysis of the analyte metal in the sample solution. The low temperature of the natural gas and air flame, compared to the other excitation methods such as arcs, sparks and rare gas plasmas, limit the method to easily ionized metals. Since the temperature is not high enough to excite transition metals, the method is selective toward detection of alkali and alkaline earth metals. On the other hand the low temperatures make this method susceptible to certain disadvantages, most of them related to interference and the stability of the flame and aspiration conditions. Fuel and oxidant flow rates and purity, aspiration rates, solution viscosity, contaminants in the samples etc. affect these. It is therefore very important to measure the emission of the standard and unknown solutions under conditions that are as nearly identical as possible.

Instrument

The instrument consists of six parts namely,

i. Burner
ii. Atomizer which dispenses the solution as a fine spray in to the flame
iii. Filter system or diffraction grating, which isolates the specific wavelength of emitted light which is characteristics of the sample element
iv. Photocell which converts the light energy in to electric current
v. Amplifier which amplifies the feeble current
vi. Galvanometer which measures electric signal

Determination of Sodium and Potassium

Reagents

i. Sodium stock solution: 0.254 g NaCl dissolved in 100 mL deionized distilled water gives 1000 ppm sodium.
ii. Potassium stock solution:0.191g KCl dissolved in 100 mL of deionized distilled water gives 1000 ppm potassium.
iii. Combined working standard solution of sodium and potassium ions. This solution has a good similarity for sodium and potassium with regard to their concentration generally found in milk and compensate the interference of one element in the estimation other. The concentration of working solution should be between 5 and 10 ppm.

Preparation of Sample

The milk/skim milk sample is diluted with deionized/distilled water so that the metal concentration in the diluted sample is in the range of 5–8 ppm.

Method

Connect the instrument to the air compressor and gas cylinder. Switch on the air compressor and adjust air pressure through the instrument. Open outlet of the gas cylinder and ignite the burner. Standardize the instrument with the standard solutions after setting the readout to zero using deionized/ distilled water. Check for accuracy and reproducibility by measuring the standard several times. Be sure to aspirate deionized/distilled water between measurements. Feed the diluted milk sample in to the instrument and record the reading in ppm.

Calculation

Sodium content in milk = reading in ppm × dilution factor

Potassium content in milk = reading in ppm×dilution factor

Precaution

i. All solutions should be free from any impurities
ii. Avoid use of thumb while mixing solutions.Use glass spatula.
iii. Always use deionized or double distilled water.
iv. Pressure of air and burning gas should remain constant throughout the experiment

Membrane Separation Process

Membrane separation processes operate without heating and therefore use less energy and therefore use less energy than conventional thermal separation process such as distillation, sublimation, or crystallization. Furthermore, using membranes enables separations to take place that would be impossible using thermal separation methods. Particles are separated on the basis of their molecular size and shape with the use of pressure and specifically designed semi-permeable membranes. Depending on the type of membrane, the selective separation of certain individual substances or substance mixtures is possible. Important technical applications include the production of drinking water by reverse osmosis ,filtrations in the food industry, the recovery of organic vapours such as petro-chemical vapour recovery and the electrolysis for chlorine production. There are different types of membrane process which are used as analytical techniques in isolation and purification of analyte.

Principle

When a solution and water are separated by a semi permeable membrane, the water will move in to the solution to equilibrate the system. This is known as osmotic pressure. The water is forced to move down the concentration gradient i.e. from low to high concentration if a mechanical force is applied to exceed the osmotic pressure. Permeates designates the liquid passing through the membrane and retentates designates the fraction not passing through the membrane. Depending on the mode of operation and types of membrane used, the different types of membrane separation processes are:

i. Microfiltration.
ii. Ultrafiltration.
iii. Nanofiltration.
iv. Reverse osmosis.

Pore Size and Selectivity of Membrane

Filter membranes of membrane separation process are divided into four classes according to their pore size

Pore size	Molecular mass	Process	Filtration pressure	Items removed
>0.1 µm	>5000 kDa	Microfiltration	<2 bar	Large bacteria, yeast, particles
110–2 nm	5–5000 kDa	Ultrafiltration	1–1 bar	Bacteria, macromolecules, proteins, larger viruses
2–1 nm	0.1–5 kDa	Nanofiltration	3–20 bar	Viruses, divalent ions
<1 nm	<100 Da	Reverse Osmosis	10–80 bar	Salts, small organic molecules

Fractionation of Milk Protein by Ultrafiltration

Principle

Ultrafiltration selectively separates macromolecules/particles/molecules on the basis of differences in their size where in water and other small molecules are driven through a semi permeable membrane by a trans membrane force such as centrifugation or pressure. The membrane pores range in diameter from 1 to 20 nm, the diameter is chosen such that the protein/peptide of interest is too large to pass through. The process employs membranes with molecular cut off limits from 500 to 100,000 Daltons and employing operating pressure from 10–200 psi. Several types of apparatus are available for UF on a small scale using centrifugal force to suck and derive the small molecules through semi permeable membrane. Stirred cells are easy and convenient to use at laboratory scale for small volumes of 1–400 mL. UF membranes are generally made up of regenerated cellulose, cellulose acetate, polycarbonate, polysulfone, polyamide polyvinyl chloride and polyvinyl di fluoride. There are few inorganic based membranes having a support layer of carbon with an ultrafiltration layer of zirconium oxide, with extremely good chemical and thermal resistance. These types of membrane can be operated at 100 °C and sterilizable by autoclaving.

Fractionation of milk proteins in to different constituents are being possible now due to invention of different membrane composition, configuration, and porosity. Milk proteins are theoretically separable directly by selective membrane filtration of skim milkfrom largest size to least size. The sequence is MFGM lipoprotein>casein micelles>immunoglobulin>lactoferrin> bovine serum albumin>β lactoglobuline>α lactalbumin>milk protein derived peptides. Ultrafiltration has been extensively used for the fractionation of whey proteins and bioactive peptides for their nutritional, pharmaceutical and therapeutic use.

Materials

i. Centrifugal devices fitted with different cut off membranes.
ii. Centrifuge.
iii. Stirred cell.
iv. Nitrogen gas cylinder.
v. Magnetic stirrer.
vi. Membranes of different cut off limits.

Procedure

A. Ultrafiltration by centrifugation

i. Insert sample reservoir into filtrate vials.
ii. Add solution of sample reservoir (2 mL maximum volume). Do not touch membrane top with pipette tip. Use a covered rotor in order to minimize sample evaporation. Sealing of device by attaching the retentate vial to the sample reservoir can also be done to minimize the evaporation. The minimum length which a rotor adapter must accept is 137 mm.
iii. Place covered device and attached filtrate vial in to the centrifuge rotor. Use similar device for counter balance.
iv. Spin centrifugal devices at 4000–7500×*g* (with retentate vial in place) or 4000–6500×*g* (without retentate vial) until desired concentration is achieved.
v. Remove centrifugal filter assembly from centrifuge and separate filtrate vial from membrane support base. For filtration applications, filtrate may either be discarded or left in the filtrate vial.
vi. Place retentate vial over sample reservoir and invert unit to recover the retentate.
vii. Centrifuge at 300–1000×*g* for 2 min to transfer concentrate in to retentate vial.
viii. Remove device from centrifuge. Separate retentate vial from concentrator. Use the cap to cover retentate or filtrate vial.

B. Ultrafiltration using stirred cell

i. Pretreat the ultrafiltration membrane according to the manufacturer's instruction to remove preservatives. Sock the membrane.
ii. Assemble the stirred cell according to the manufacturer's instructions. Ensure the membrane is placed the correct way (shiny surface uppermost).

iii. Pre-filter or centrifuge the solution to remove particulate matter and gently pour in to the stirred cell. Attach the cap assembly and place in the retaining stand.

iv. Ensure the pressure relief cap is closed and attach a regulated pressure source, typically a nitrogen cylinder. Place the assembly on a magnetic stirrer.

v. Apply the minimum pressure required to give an acceptable flow rate. Higher pressure will lead to increased concentration polarization and fouling and therefore results in a reduced flux rate. Do not exceed the maximum pressure recommended by the manufacturer. Place the filtrate tubing in to a collection vessel.

vi. Turn on the magnetic stirrer and adjust the stirring rate so that the vortex is no more than one third the depth of the solution. Excess stirring will denature the protein by shear forces and foaming. Insufficient stirring will cause increased concentration polarization.

vii. When the desired concentration is achieved, turn off the pressure and open the pressure release valve. Continue stirring for 5 min to suspend the polarized layer.

viii. Remove the cap assembly and gently pour out the concentrate.

ix. The different steps as described above can be repeated using membranes of different cutoff limits for fractionation of milk proteins/peptides.

Appendix

1. Stain for Electrophoresis

Name of the Stain	Specificity of stain
Amido black 10B	Protein
8-Anilinonapthalene-1-sulphonic acid	Fluorescent stain for proteins
Coomassie brilliant blue G250	Commonly used stain for protein
Coomassie brilliant blue R250	Sensitive, durable stain for protein
Fast Green FCF	Quantitative determination of protein by color intensity
Silver stain	For all proteins including glycoproteins are highly sensitive
Sudan black	For lipoproteins

Source: Research techniques in dairy chemistry by Mann *et al.*

2. Characteristics of Some Commonly Used Ion Exchangers

Type	Matrices	Functional group
Weak cation exchanger	Agarose, cellulose, dextran,polyacrylate	Carboxyl,carboxymethyl
Strong cation exchanger	Cellulose, dextran, polystyrene	Sulpho, sulphomethyl, sulphopropyl
Weak anion exchanger	Agarose, cellulose, dextran, polystyrine	Aminoethyl, diethylaminoethyl
Strong anion exchanger	Cellulose, dextran, polystyrine	Trimethylamino methyl, Triethylaminoethyl, Dimethyl-2-hydroxyethyl aminoethyl, Ciethyl-2-hydroxypropyl aminoethyl

Source: Research techniques in dairy chemistry by Mann *et al.*

3. Common Sources of Electromagnetic Radiation for UV–VIS Spectrophotometer

Source	Wavelength region	Use
Tungsten lamp	320–2400 nm	Visible molecular absorption
H_2 and D_2 lamp	160–380 nm	UV molecular absorption

17

Physical Properties of Milk and Milk Products

The primary constituents of milk are water, fat, protein, sugar (lactose), minerals, vitamins, and enzymes. Milk is a thin emulsion made up of two phases: A continuous aqueous colloidal phase and a dispersed phase of oil or fat. It is an opaque white liquid containing fat in an emulsion, protein, and a few minerals in a colloidal suspension, and lactose along with soluble proteins and a few minerals in a true solution. The physical characteristics of milk are comparable to those of water, but they are altered by the degree of dispersion of the colloidal and emulsified components as well as the presence of different solutes, such as salts, lactose, and protein, in the continuous phase.

Physical Appearance, Color and Optical Properties

Fat globules and casein micelles scatter light, giving the appearance of turbid and opaque milk. The way that molecules scatter light has an impact on optical characteristics. When the wavelength of light and the particle's magnitude coincide, light scattering happens. Consequently, light with shorter wavelengths is scattered by smaller particles and vice versa. Because casein micelles scatter blue light more than red light at shorter wavelengths, skim milk appears slightly blue. Cow's milk has a creamy color because of beta-carotene, a carotenoid precursor to vitamin A. The greenish tinge in whey is due to the presence of riboflavin. Turbid and opaque due to scattering of light mainly by protein micelles and fat globules.

Flavor of Milk

Milk's inherent sweetness comes from the interaction of all of its ingredients. A number of factors cause off-flavors in milk to develop very quickly. The feed that animals eat can cause some unwanted flavors. Fruity, barny, malty, or acid flavors in milk are the result of bacterial growth. Enzyme activity can also result in artificial flavors; one well-known example is rancidity brought on by lipase activity. In milk, oxidative reactions can give it a cardboard taste. Cooking milk can give it a cooked flavor.

Colligative Properties

Colligative properties are those of solutions that depend on the quantity of molecules in a given amount of solvent, rather than on the characteristics or identities (such as size or mass) of the molecules, or the quantity of dissolved particles in the solution, without regard to the solute identities.

Four frequently researched colligative properties of milk are:

1. Freezing point
2. Boiling point
3. Vapour pressure
4. Osmotic pressure

These characteristics can be used to determine the molecular weight of the solute because they provide information on the quantity of solute particles in the solution.

Specific Gravity and Density

The specific gravity of a substance is the ratio of its mass at a known temperature to the mass of an equal volume of water at 4° C. whereas density is the mass per unit volume at a specific temperature. Milk is heavier than water. The average specific gravity of normal bovine milk varies from 1.029 to 1.032. The specific gravity of skim milk varies from 1.034 to 1.036. The specific gravity of different milk constituents are Fat: 0.930, Protein: 1.346, Lactose: 1.666, Salts: 4.120, and Water: 1.000. Though specific gravity varies with temperature, (lower at higher temperatures and vice versa), the rate of this variation is not uniform.

Significance

Specific gravity and density are used in various applications such as quality control, density determination, and fluid mechanics. Understanding milk density helps in calculating its mass, assessing nutritional content, and ensuring proper processing and packaging.

Factors affecting

The specific gravity and density of milk are affected by the following factors

1. Temperature
2. Breed of animal
3. Composition of milk

A temperature increase causes a significant decrease in specific gravity/density. Since the thermal expansion coefficient controls how temperature affects specific gravity and density, temperature must be mentioned when talking about specific gravity and density.

Breed-specific differences exist in milk's specific gravity and density. The specific gravity of Ayrshire cows is marginally lower than that of Jersey and Holstein cows.

The season, the kind of feed, and the method of milking all affect the composition of milk. The specific gravity and density of milk are impacted by variations in its composition.

Recknagel Phenomenon: The density or specific gravity of milk just after milking is lower than the same milk held for long periods, especially milk under cold storage. Such phenomenon is known as the Reckngel phenomenon. The increase in the hydration of the protein at low temperatures may be the major cause for such a phenomenon rather than the escape of the air bubbles. Subsequent work carried out by other scientists attributed this phenomenon to the ratio between the liquid and solid.

Viscosity

The resistance of a fluid to flow is measured by its viscosity. The primary cause of milk's viscosity is the fat and milk proteins that are present as a colloidal system; lactose and salts in solution with water play a minor role as well. The relative viscosity of cow milk, buffalo milk skim milk, and whey is 2.00 cp, 1.80 cp, 1.50 cp, and 1.20 cp, respectively at 20° C.

Significance

Understanding milk viscosity is crucial for optimizing processing conditions, ensuring product consistency, and developing new dairy products.

Factors affecting

Milk viscosity depends on factors which affect its texture and processing characteristics.

1. State and concentration of protein
2. State and concentration of fat
3. Temperature of milk
4. Age of milk
5. Processing operation such as agitation and homogenization

Protein concentration, particle number and size and degree of hydration affect the viscosity of milk. Fat globule size and the extent of fat clustering influence the viscosity of milk. Increase of temperature causes a marked reduction of viscosity. The viscosity of pasteurized milk, both homogenized and unhomogenized, increases with age. Cooler temperatures increase viscosity due to the increased voluminosity of casein micelles whereas temperatures above 65°C increase viscosity due to the denaturation of whey proteinsThe casein micelle voluminosity of milk increases in response to changes in its pH. Agitation affects viscosity in different ways. Agitation can sometimes disperse fat globules that have undergone cold agglutination, resulting in a decrease in viscosity, or it can cause partial coalescence of the fat globules, increasing the viscosity. Homogenization increases viscosity. During homogenization fat globules are greatly sub-divided and this contributes to a higher viscosity of milk.

Surface tension

Surface tension is the force acting on the surface of a liquid, tending to minimize its surface area.

The surface activity of milk is related to proteins, fat, phospholipids and fresh fatty acids present in it. Homogenization and heat sterilization increase the surface tension of milk. The surface tension between milk and air has been reported to range between 40 and 60 dynes/cm, with an average of 52 dynes/cm at 20° C.

Significance

It affects milk's behaviour in processes like foaming, emulsification and droplet formation, impacting product quality and processing efficiency.

Factors affecting

1. Milk composition
2. Heat treatment
3. Processing operation
4. Lipolysis

Proteins, phospholipids, mono- and diglycerides, and free fatty acid salts are the main surfactants found in milk. Compared to other milk proteins, immunoglobulins are less effective surfactants. Lactose and salts don't really add much to the interfacial tension in milk. Lipids and proteins lower milk's surface tension. Surface tension can be effectively reduced by whey proteins. Low surface tension is also caused by the membrane of fat globules that are released from fat globules. With the exception of sterilized milk, which exhibits higher surface tension due to protein denaturation, heat-treating milk has little effect on its surface tension.

When raw milk is homogenized, surface-active fatty acids are released and lipolysis by the natural milk lipase is triggered, which lowers surface tension. Surface tension slightly increases when pasteurized milk is homogenized. The surface tension of milk is not significantly affected by pasteurization.

Boiling Point

Milk has a slightly higher boiling point than pure water because of the dissolved solids. This elevation occurs because the dissolved solutes disrupt the formation of vapor bubbles, requiring a higher temperature to reach the boiling point.

The boiling point of water at atmospheric pressure is 100.15° C.

Significance

It is important in dairy processing to ensure proper pasteurization and sterilization. Boiling point determination helps in the detection of adulteration in milk.

Factors Affecting

1. Composition of milk
2. Pressure

The ingredients in milk raise the boiling point of the liquid. When water is added to milk, the concentration of the dissolved substances that raise the boiling point of the milk is lowered, causing adulterated milk to boil at a lower temperature than regular milk. The boiling point of milk rises with an increase in pressure.

Freezing Point

The freezing point of milk is lower than that of water due to the presence of dissolved solutes, primarily lactose and minerals. The freezing point of bovine milk is usually in the range of –0.512° C to –0.550° C, with a mean value close to –0.522° C.

Significance

The solutes present in milk interfere with the formation of ice crystals, requiring a lower temperature for freezing to occur. Understanding the freezing point is crucial for determining storage conditions and preventing quality deterioration during freezing.

Factors affecting

1. Composition of milk
2. Processing operations

The major components that affect the freezing point of milk are lactose and soluble salts. Lactose and chloride represent 75 to 85% of the entire freezing point depression of milk. Fat, protein, and colloidal calcium phosphate have negligible effect on the freezing point of milk because of their high molecular weight.

Variations in the freezing point of milk have been attributed to seasonality, feed, stage of lactation, water intake, and breed of cow, heat stress, and time of the day. These factors have little influence on the freezing point of milk. Direct UHT treatment involves the addition of water. The additional water is removed during flush cooling, which also removes gases from milk, causing a small increase in freezing point. Vacuum treatment of milk has been shown to increase the freezing point of milk. Fermentation of milk has an effect on the freezing point of milk as fermentation of one mole of lactose results in the formation of four moles of lactic acid.

Redox Potential

An electron is transferred from an electron donor (reducing agent) to an electron acceptor (oxidizing agent) in oxidation-reduction (redox) reactions. It is said that an electron-losing species is oxidized, whereas an electron-accepting species is reduced. Redox reactions must be coupled, and an oxidation reaction and a reduction reaction happen at the same time because there can be no net transfer of electrons into or out of a system. A solution's redox potential, represented by the symbol E_h, is the half-cell's potential at the inert electrode expressed in volts.

The E_h of milk is usually in the range +0.25 to +0.35 V at 25 °C, at pH 6.6–6.7, and in equilibrium with air.

Significance

The redox potential plays a major role in the measurement of the degree of microbial contamination of milk and its shelf-life and its suitability for storage, oxidative deterioration and other processing operations.

Factors affecting

1. Concentration of dissolved oxygen
2. Concentration of ascorbic acid
3. Exposure to light

4. Processing operations
5. Fermentation
6. pH

The main factor influencing milk's redox potential is its dissolved oxygen concentration. The redox potential of fresh milk ranges from +0.20 to +0.30 volts, primarily because of the dissolved oxygen. Milk loses its oxygen content when heated, and the production of -SH groups from β-lactoglobulin also lowers milk's redox potential.

The ascorbic acid content of milk (11.2–17.2 mg/L) is sufficient enough to affect the milk's redox potential. All of the ascorbic acid in freshly drawn milk is in the reduced form, but it can oxidize reversibly to become dehydroascorbate. The redox potential is stabilized by the ascorbate/dehydroascorbate in milk.

The redox potential of milk is influenced by the exposure to light and by anumber of processing operations which changes the concentration of O_2 in the milk. Heating of milk causes a decrease in redox potential mainly due to the denaturation of β-lactoglobulin and loss of O_2. Maillard reaction products can also influence the redox potential of heated milk products, especially in dried milk products.

The process of lactose fermentation in milk during the microorganisms' growth significantly impacts the milk's redox potential. Once the bacteria have consumed all of the available O_2, the redox potential rapidly decreases. Cheese and other fermented milk products have a negative redox potential as a result.

Redox potential fluctuated within the typical range of milk pH, by +0.03 volts for each unit drop in pH.

Acidity and pH

The pH of freshly drawn milk ranges from 6.5 to 6.7 at 25^0 C, and its titratable acid content, measured as lactic acid, is between 0.14 and 0.18%. Freshly drawn milk does not have developed acidity; instead, its slightly lower pH than neutral is explained by the presence of casein, citrate, carbon dioxide, and other substances. Lactic acid, which is produced when bacteria act on the lactose in milk, is the cause of developed acidity.

Significance

pH controls the enzyme activity, coagulation process of milk, growth of bacteria and cultures.

Factors affecting

1. Composition of milk
2. Species and breed of animal
3. Processing operations

Natural acidity in milk increases with increasing solid-not-fat (SNF) content and vice versa. The composition of milk varies between species and breeds, which could lead to a slight variation in the pH of the milk. Processing activities in milk have an impact on the acid-base equilibrium. Because of the CO_2 loss and calcium phosphate precipitation during pasteurisation, there is a slight pH change. When heat treatments surpass 100^0 Celsius, lactose breaks down into different organic acids, particularly formic acid, which lowers pH.Since the formation of ice crystals during slow freezing concentrates the solutes in the milk's aqueous phase, precipitating calcium phosphate and simultaneously releasing H+, slow freezing of milk lowers its pH. When milk is concentrated by evaporating water, more colloidal calcium phosphate is formed and the solubility of calcium phosphate is exceeded. This lowers the pH of the milk.

Buffering capacity of milk

Due to the presence of CO_2, protein, phosphate, citrate, and several other minor components, fresh milk functions as a complex buffer. Lactate and other organic anions are introduced by bacterial action as extra buffers.

The ability of a milk solution to withstand pH changes is known as its buffer capacity.

1. Even if a tiny bit of acid is added to the milk, the pH stays the same.
2. The acidity will be balanced by the salt and milk proteins.
3. Phosphate and casein are crucial elements of milk buffer.

Osmotic Pressure

The lowest pressure needed in a solution to stop solvent molecules from passing through a semi-permeable membrane is known as the osmotic pressure. Osmotic pressure is a good indicator of milk's solute concentration.

The osmotic pressure of milk is 700 k Pa.

Significance

It stands for the amount of force required to stop water from passing through a semipermeable barrier. This pressure in milk is caused by dissolved materials such as proteins, minerals, and lactose. Increased osmotic pressure is a sign of increased solute concentration, which impacts a number of attributes including taste, texture, and microbiological stability.

Factors affecting

i. Concentration of plasma protein especially albumin

ii. Temperature

iii. Surface area

Albumin's large molecular weight is thought to be responsible for most of its osmotic effect, with its negative charge thought to be responsible for the remainder. The latter allows albumin to attract positively charged molecules and, ultimately, water. By promoting the movement and dissociation of solute particles and thereby raising their effective concentration, raising the temperature of a solution can increase the osmotic pressure. The molecules can move more freely across a larger surface area; conversely, slower and more restricted movements occur when the surface area is smaller.

Water activity

Water activity (a_w) measures the amount of water available for biological and chemical reactions. It is defined as the ratio of vapour pressure of water in food (p) to vapour pressure of pure water (p_0) at the same temperature.

$$a_w = p / p_o$$

The water activity of milk is 0.993 at 25 ^{0}C.

Significance

The water activity of most food is higher than 0.95, which encourages the growth of mould, yeast, and bacteria. It's critical to understand a food's water activity when creating a Hazard Analysis Critical Control (HACCP) plan. For many products, determining the water activity of an ingredient or product is essential when performing a hazard analysis.

Factors affecting

i. Temperature

ii. Solutes

iii. Drying and freezing

Water activity is highly dependent on temperature. Because of variations in water binding, water dissociation, solute solubility in water, and matrix state, temperature affects water activity. While solute solubility may play a role, matrix state is typically the primary source of control. The addition of solutes, like sugar or salt, reduces the activity of water. Removing water physically by drying reduces water activity. Water activity of milk is decreased by freezing.

Refractive Index

The refractive index measures how light bends as it passes through a substance. The refractive index of milk is difficult to estimate due to light scattering by fat globules and casein micelles.

The refractive index of milk is 1.3440 – 1.3485 at 20^0 C.

Significance

This property is utilized in quality control to assess milk composition, detect adulteration and monitor processing conditions.

Factors affecting

i. Composition of milk
ii. Temperature

Protein is the solid component of milk that contributes the most to the refractive index of milk serum, followed by lactose. Since fat and calcium caseinate obstruct the flow of light through milk, they are precipitated out of the milk before its refractive index is calculated. The refractive index of milk is currently determined using an Abbe refractometer.

Values for the refractive index are typically calculated at room temperature. As the temperature rises, the liquid loses density and viscosity, allowing light to move through the medium more quickly. Because of the smaller ratio, this leads to a lower value for the refractive index.

Specific Conductance

A substance's specific conductance indicates how well it can conduct electricity. It shows ions and dissolved solids in milk, indicating the composition and quality of the milk.

The specific conductance of milk is 0.0040 – 0.0050 ohm^{-1} cm^{-1}

Significance

A rapid method to identify subclinical mastitis in milk is to measure its specific conductance.

Factors affecting

i. Temperature
ii. Fat content
iii. Acidity
iv. Dilution
v. Concentration
vi. Neutralisation of milk

An increase in temperature causes the electrolytes to dissociate more readily and the viscosity to decrease, increasing electrical conductivity. As milk fat content rises, its electrical conductivity falls because fat globules obstruct ion mobility and the fat phase acts as an inert diluent. Compared to whole milk, skim milk has a higher electrical conductivity. As acidity rises, colloidal minerals become more soluble, leading to an increase in electrical conductivity. When milk is diluted, the concentration of electrolytes decreases, which is somewhat offset by an increase in electrolyte dissociation. This results in a decrease in electrical conductivity. When neutralizers are added to milk, the concentration of electrolytes rises, which in turn increases electrical conductivity.

Ionic Strength

The concentration of ions in a solution is known as its ionic strength. It affects the activity of ions.

The ionic strength of a solution is expressed by the following relation

$I = \frac{1}{2} \Sigma c_i z_i^2$

Where ci is the molar concentration of the ion i and z_i is its charge.

Ionic conductance of milk is 0.08 M.

Significance

The chemical properties and interactions of milk are influenced by the concentration of ions in a solution, which is measured by ionic strength. It displays the quantity of charged particles, such as phosphate, potassium, and calcium ions, in milk.

Thermal Conductivity

The capacity of a material to conduct heat is measured by its thermal conductivity.

Thermal conductivity of whole milk (2.9% fat) is 0.559 W m^{-1} K^{-1}

Significance

It is essential for understanding heat transfer during processing, pasteurization, and storage, ensuring product safety and quality.

Factors affecting

i. Fat content

ii. Temperature

While milk and cream's thermal conductivity rises with temperature, it falls with an increase in total solids or fat content, especially at higher temperatures.

Because of variations in the quantity of air trapped in the powder, dried milk products' bulk density affects their thermal conductivity in addition to their composition.

Thermal Diffusivity

Thermal diffusivity quantifies how quickly heat can move through a substance. Thermal diffusivity of milk at 20^0 C is $1.25 * 10^{-7} m^2 s^{-1}$

Significance

Comprehending thermal diffusivity is crucial for maximising processing parameters, averting over- or undercooking, and ensuring the safety and quality of the final product.

Factors affecting

i. Composition and temperature
ii. Heat transfer rates during heating and cooling processes

References

De, Sukumar. 2004. "Outlines of Dairy Technology". Oxford University Press. New Delhi, India.

Fox, PF, Uniacke-Lowe, T, McSweeney PLH, O'Mahony JA. 2015 "Dairy Chemistry and Biochemistry, 2nd Edition. Springer.

Jenness R and Patton S. 1959. "Principles of Dairy Chemistry". John Wiley & Sons, Inc. New York

Mathur MP, Roy DD, Dinakar P. 2019. "Textbook of Dairy Chemistry". Indian Council of Agricultural Research, New Delhi.

18

Vitamins in Milk and Milk Products

Vitamins are naturally occurring, vital nutrients that are needed in small amounts. They are important for bone and tissue health, wound healing, growth and development, immune system function, and other biological processes. These vital organic substances serve a variety of biochemical purposes. Vitamins, like minerals, are not biosynthesized by the body. Consequently, in order to maintain our health, we must obtain them from the food we eat or, in rare circumstances, supplements.

Vitamins in Milk

Together with water-soluble vitamins like vitamin C and B complex and provitamins, milk also contains a variety of fat-soluble vitamins like A, D, E, and K. Some of these vitamins do, however, have very low concentrations. The concentrations and type of vitamins present in milk are given in Table 1.

Table 1: Concentration of the vitamins in milk

Vitamins	**Class**	**Concentration/L**
Vitamin A	Fat soluble	1590 I.U.*
Vitamin D	Fat soluble	22.1 I.U.
Vitamin E	Fat soluble	1.0 mg
Vitamin K	Fat soluble	0.04 mg
Vitamin B_1	Water soluble	0.40 mg
Vitamin B_2	Water soluble	1.5 mg
Pantothenic acid	Water soluble	3.0 mg
Biotin	Water soluble	50.0 μg
Niacin	Water soluble	0.2-1.2 mg
Pyridoxine B_6	Water soluble	0.70 μg
Folic acid	Water soluble	1.0 μg
Vitamin B_{12}	Water soluble	7.0 μg
Vitamin C	Water soluble	20.0 mg
Inositol	Water soluble	180 mg
Choline	Water soluble	150 mg

* A Retinol Equivalent (RE) is equal to 3.33 I.U. retinol or10 I.U. β-carotene

Ref: Textbook of Dairy Chemistry by Mathur *et.al.* 2005.

On the basis of their solubility vitamins are distinguished into two

1. Fat soluble vitamins A, D, E and K,
2. Water soluble vitamins, the B-complex and vitamin C (Ascorbic acid).

Fat Soluble vitamins

Vitamin A (Retinol)

Plant foods contain carotenoids, which are the precursors of vitamin A, whereas animal foods provide preformed vitamin A in the form of retinyl esters.

Structure

Vitamin A in animal food is present in two forms A_1 and A_2 (Fig. 1)

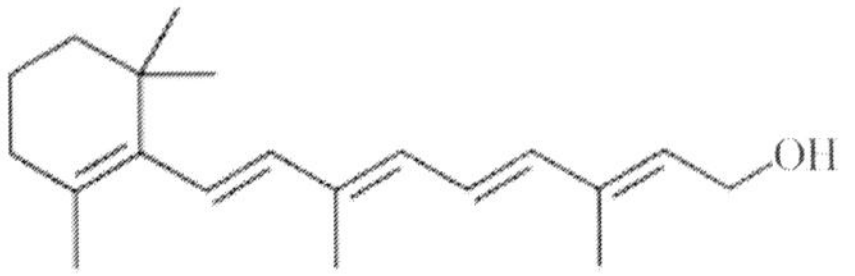

OH

Vitamin A1

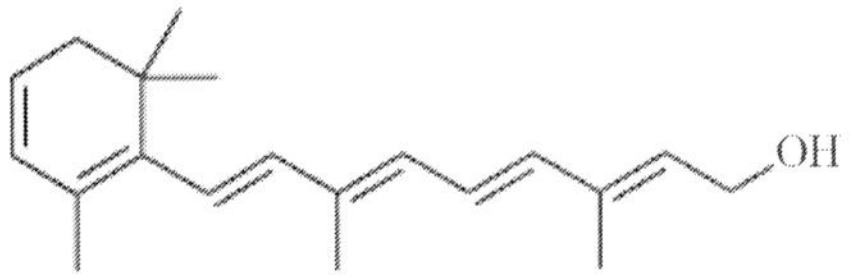

OH

Vitamin A2

Fig. 1: The chemical structure of vitamin A_1 and A_2.

Dietary source

The main dietary sources of retinol are liver, eggs, and dairy products. Carrots, pumpkin, sweet potatoes, and other deep orange fruits and vegetables are good providers of β-carotene, as are spinach and other dark green leafy vegetables.

Biological role

1. Has a significant part in the process of normal vision
2. Vital for the development and upkeep of epithelial tissue, the skins outermost layer of cells
3. Necessary for healthy growth and reproduction

Deficiency syndrome

Night blindness, xeropthalmia (a progressive form of blindness brought on by drying out the cornea of the eye), and keratinization (a build-up of keratin in tissues of the digestive, respiratory, and urogenital tracts) are all consequences of vitamin A deficiency.

Activity in milk

Vitamin A activity is present in milk as retinol, retinyl esters and as carotenes. An average 100g of whole cows' milk has 21 μg of carotene and 52 μg of retinol. Retinol levels in human milk and colostrum are typically 58 and 155 μg/100g, respectively. The carotenoids in milk are what give it its color in addition to serving as a provitamin A.

Factors Affecting

Season, diets of the animal, breed, lactation period and processing operations are the factors that affect vitamin A content of milk and milk products.

Vitamin D (Calciferol)

In contrast to other vitamins, when exposed to sunlight, the skin can produce cholecalciferol, or vitamin D3, from the steroid precursor 7-dehydrocholesterol. It is not necessary to consume preformed vitamin D from the diet if one receives enough sun exposure.

Structure

Vitamin-D occurs in several chemical forms (at least 6 forms) of which two most important are vitamin-D_2 and vitamin-D_3 (Fig. 2). Two principal forms of vitamin D are:

1. Ergocalciferol (vitamin-D_2)
2. Cholecalciferol (vitamin-D_3)

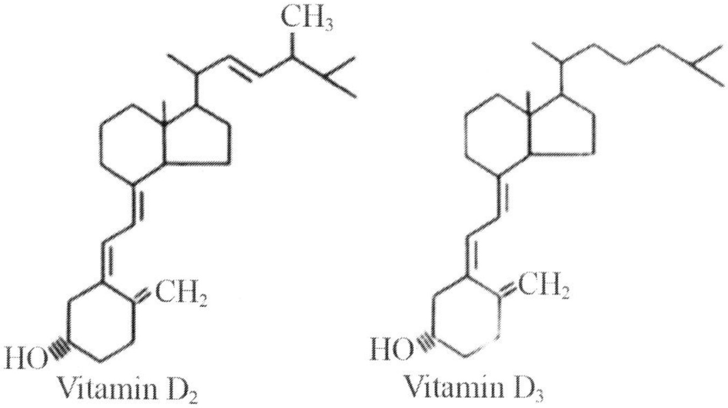

Fig. 2: The chemical structure of vitamin D_2 and D_3.

Dietary source

Very few foods have a substantial amount of vitamin D in them. The main source of vitamin D, aside from the body's in-situ conversion, is animal-based food, such as egg yolks, fatty fish, and liver. Cow's milk without fortification is not a significant source of vitamin D.

Biological role

1. Increased absorption of calcium and phosphorus from intestine. The retention of these minerals by kidney and the transfer from bone to blood are also promoted by vitamin D
2. Better utilization of Ca & P which ultimately helps in normal bone formation
3. Normal teeth formation

Deficiency Syndrome

1. Ricket is the classic syndrome of vitamin D deficiency which is characterized by insufficient mineralization of bone, leading to skeletal abnormalities and growth retardation
2. Osteomalacia in older people

Activity in Milk

The primary form of vitamin D found in human and cow milk is 25 (OH) D3. The majority of the vitamin D in breastfed infants' blood serum is reportedly attributed to this compound. One litre of milk provides only 10–20% of the recommended daily allowance for vitamin D, with whole cows' milk containing only 0.03 µg per 100 g. Dairy products with added vitamin D, such as fortified milk, are a significant dietary source; however, unfortified milk products have relatively little vitamin D. The amount of vitamin D in milk varies depending on sunlight exposure.

Factors Affecting

Breed of the animal, season, exposure to sunlight, season, heat and storage are the factors that affect the vitamin D content of milk.

Vitamin E (Tocopherol)

Vitamin E is a fat-soluble vitamin present in food with multiple forms. Vitamin E activity in food can be expressed as tocopherol equivalents (TE).

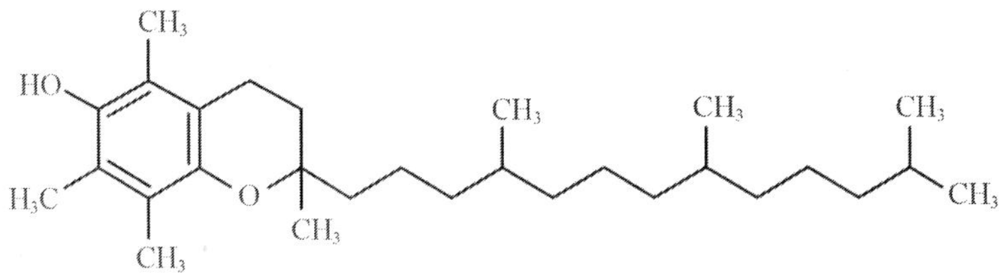

Fig. 3: The chemical structure of DL- α- tocopherol.

Structure

α- tocopherol (Fig. 3) is the only form of vitamin E that the human body uses.

Dietary Source

The main sources of vitamin E include liver, egg yolks, nuts, seeds, green and leafy vegetables, wheat germ, whole-grain cereal products, and polyunsaturated vegetable oils and their derivatives.

Biological Role

Because vitamin E has the ability to act as an antioxidant, it has various biological and physiological functions in the body.

1. Stops the oxidation of unsaturated fatty acids in bodily tissues, which stops harmful lipo-peroxides from forming. Vitamin E plays particularly a significant role in the lungs, where cells are most exposed to oxygen
2. Stops vitamin A from oxidising or from being destroyed by oxidation
3. Exerts a protective effect on red and white blood cell

Deficiency Syndrome

A vitamin E deficiency is rare in humans because it can be found in many different foods and supplements. A vitamin E deficiency can occur in people with digestive disorders or conditions where the body does not properly absorb fat, such as pancreatitis, cystic fibrosis, or celiac disease. Some typical indicators of a deficiency are as follows:

1. Retinopathy, or damage to the eyes' retina that may impede vision,
2. Damage to the peripheral nerves, typically in the hands or feet, resulting in Pain or weakness, is known as peripheral neuropathy
3. The inability to control bodily movements and reduced immune function

Activity in Milk

Cows' milk has a very low concentration of vitamin E (0.09 mg/100g), with a higher concentration in the summer than in the winter. The concentrations in colostrum and human milk are slightly higher (~0.3 and ~1.3 mg/100g, respectively). The majority of dairy products are not significant sources of vitamin E because they have low levels of this nutrient. Nonetheless, dairy products that have been supplemented with vegetable fat have higher levels.

Factors Affecting

Breed of the animals, season and diet, lactation period, storage and light and air are the factors that affect the vitamin E content in milk.

Vitamin K

The human body needs the fat-soluble vitamin K for the post-synthesis modification of several proteins necessary for blood coagulation.

Structure

Vitamin K exists naturally in two forms: Phylloquinone (vitamin K_1) occurs only in plants and menaquinones (vitamin K_2) is synthesized only by bacteria inhibited in the gastrointestinal tract. Menadione (vitamin K_3) is a synthetic compound with vitamin K activity.

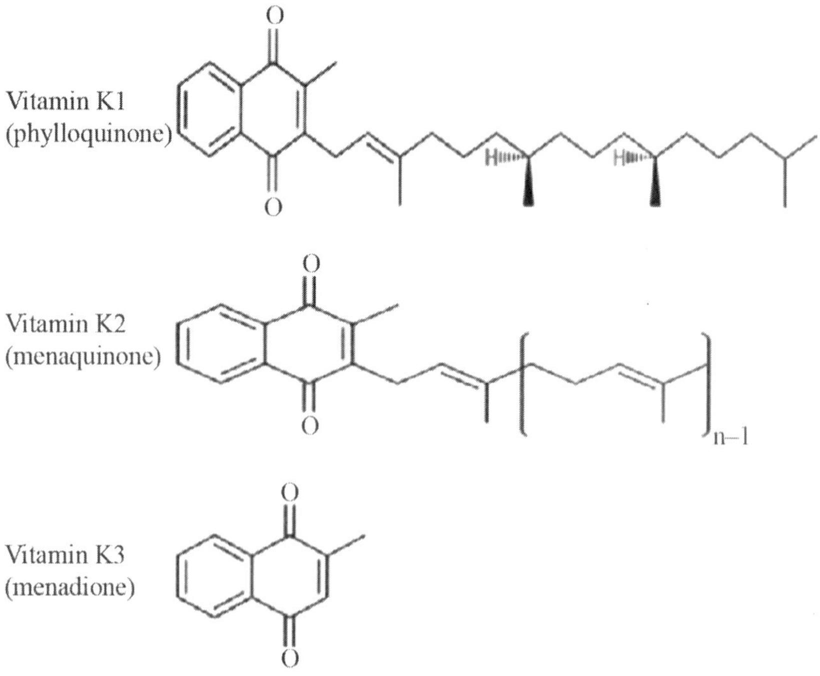

Vitamin K1 (phylloquinone)

Vitamin K2 (menaquinone)

Vitamin K3 (menadione)

Fig. 4: The chemical structures of vitamin K.

Dietary source

Green leafy vegetables, milk, and liver are the main food sources of vitamin K.

Biological Role

1. Vitamin K catalyzes the synthesis of prothrombin.
2. It activates prothrombin.
3. Activates other liver factors.
4. Plays a role in the synthesis of osteocalcin (a protein) in bone.

Deficiency Syndrome

Although rare, low vitamin K can arise from poor fat absorption. When chemical agents like antibiotics destroy the intestinal flora, the body's levels of vitamin K are also decreased.

Activity in Milk

Human milk contains approximately 0.2 μg of vitamin K per 100g, whereas whole cow's milk contains 0.4–1.8 μg. Human colostrum has a higher concentration of vitamin K, which is essential because it takes time for bacteria to synthesize vitamin K in a newborn's gut.

Factors Affecting

Feed, heat, and processing operations are the factors that affect vitamin K content in the milk and milk products.

Water Soluble Vitamins

Water soluble vitamins in milk are mainly B group vitamins and vitamin C (Table 1). The water-soluble vitamins that make up the B-group are a diverse set. Thiamine, riboflavin, niacin, biotin, pantothenic acid, pyridoxine, folic acid, and cobalamine are the vitamins that belong to the B group.

Thiamine

One of the water-soluble B vitamins is thiamine or thiamine. Another name for it is vitamin B1. In addition to being added to certain food products, thiamine can be found naturally in some foods and as a dietary supplement.

Fatigue, a swollen throat, blurred vision, and depression can all be signs of deficiency. In addition to reproductive problems, there may be liver degeneration, hair loss, and hyperemia and edema around the throat.

Activity in Milk

Milk is a good source of riboflavin. Whole milk contains about 0.17mg/100g. The majority of the riboflavin in milk (65–95%) is found in the free form; the remaining portion is found as FAD (flavin adenine dinucleotide) or FMN (flavin mononuclotide). 100g of human milk contains 0.03 mg. Additionally, dairy products are a significant source of riboflavin. Yoghurt has roughly 0.3 mg/100g, and cheese has 0.3–0.5 mg/100g.

Factors affecting

Breed of cow and seasonal variation are the two factors that affect the riboflavin content of milk.

Folic Acid

One of the B vitamins is folic acid, sometimes referred to as folacin and vitamin B_9. Because it is more stable during processing and storage, manufactured folic acid, which the body converts into folate, is used as a dietary supplement and in food fortification.

Structure

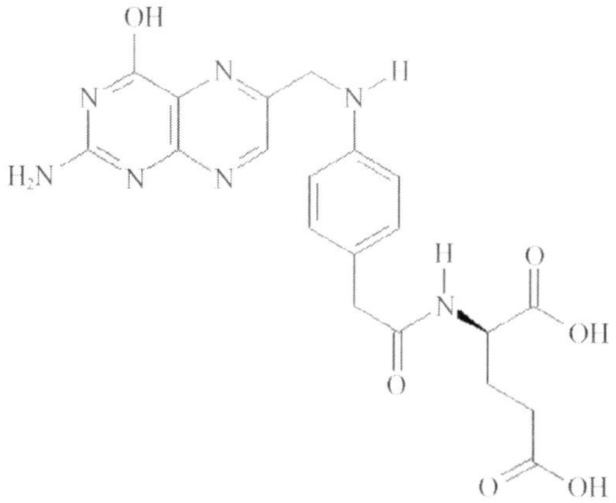

Fig. 7: The chemical structure of folic acid.

Folic acid consists of a substituted pteridine ring linked through a methylene bridge to p-aminobenzoic acid and glutamic acid.

Dietary source

Legumes, seeds, and leafy green vegetables are rich dietary sources of folic acid.

Biological role

1. Essential for the biosynthesis of nucleic acid
2. Essential for normal fat metabolism
3. Involved in the metabolism of single-carbon fragments
4. Folic acid along with inositolseem to be effective in preventing the growth of tumors
5. Helps the body to make healthy red blood cell

Deficiency Syndrome

Protein synthesis and cell division are hampered by folic acid deficiency. Megaloblastic anaemia, immune system suppression, glossitis, heartburn, constipation, diarrhoea, and nervous system issues (depression, exhaustion, mental disorientation) are among the symptoms.

Activity in Milk

Milk has approximately 6 µg of folic acid per 100g. 5-methyl-H_4 folate is the predominant form of folic acid found in milk. Forty percent of the folic acid in milk is found in conjugated polyglutamate form and is primarily bound to folate-binding proteins. Human milk's folic acid content rises from 2 to 5 µg/100g as colostrum transforms into mature milk. Higher level of folic is observed in mould ripened cheese varieties such as camembert (100 µg/100g). This is mainly due to the biosynthesis of folic acid by the moulds. Yoghurt also contains higher level of folic acid (18 µg/100g) due to the biosynthesis of folic acid by *Streptococcus thermophilus*.

Factors affecting

Light and heat treatments are the two major factors that affect the folic acid content of milk and milk products.

Pyridoxine

The body needs pyridoxine, or vitamin B6, to produce red blood cells, use the energy found in food, and maintain healthy nerve function.

Structure

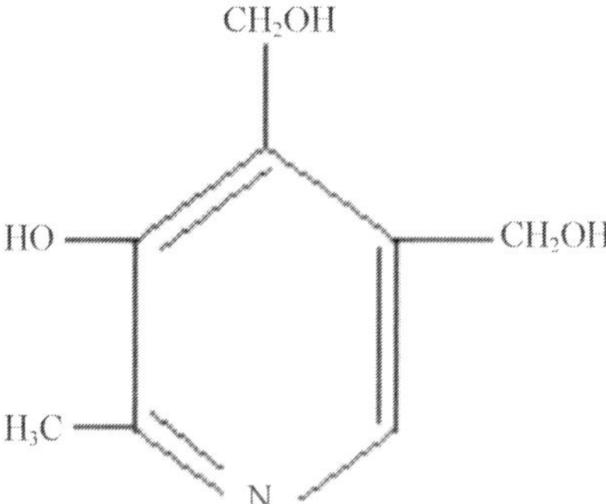

Fig. 8: The chemical structure of pyridoxine.

Dietary Source

Important sources of pyridoxine include green leafy vegetables, meat fish and poultry, shell fish, legumes, fruits and whole grains.

Biological Role

1. Pyridoxine, or vitamin B6, is a coenzyme that is essential for the metabolism of fat and protein
2. Aids in the synthesis and break down of amino acids as a coenzyme
3. Aids in the synthesis of unsaturated fatty acids from essential fatty acids
4. Acts on removal of CO_2, transfer of NH_2, removal of H_2S and other amino acids

Deficiency Syndrome

Vitamin B6 deficiency is characterized by weakness, irritability, and insomnia. Later on, it can also cause convulsions and impair immune response, growth, and motor abilities.

Activity in Milk

The average amount of vitamin B_6 in 100g of whole milk is 0.06 mg, primarily as pyridoxal (80%). With a small amount of pyridoxamine phosphate, pyridoxamine (20%) makes up the majority of the balance. Vitamin B_6 levels in colostrum are lower than in mature milk. The amount of B_6 in mature human milk is approximately 0.01 mg/100g. Dairy products are generally not a significant source of vitamin B_6. The concentration in cheese and related products ranges from approximately 0.04 mg/100g for cream cheese to 0.22

mg/100g for camembert. 100g of whole milk yoghurt has about 0.01 mg of B_6. Skim milk powder has a concentration of about 0.6 mg/100g.

Factors Affecting

Lactation period and seasonal variation influence the vitamin B_6 content of milk.

Ascorbic Acid (Vitamin C)

Ascorbic acid can be produced by most species from D-glucose or D-galactose, with the exception of primates, some birds, guinea pigs, and Indian fruit bats. Citrus fruits, as well as other fruits, berries, and vegetables, contain this water-soluble vitamin.

Structure

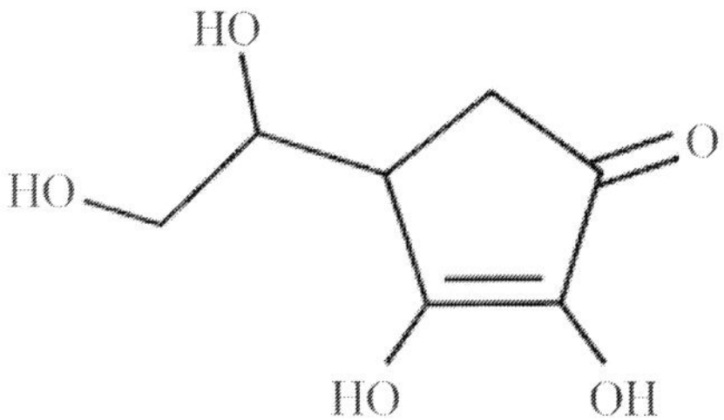

Fig. 9: The chemical structure of ascorbic acid.

Dietary Source

The richest sources of ascorbic acid are fruits and vegetables. Milk is a poor source.

Biological Role

1. Acts as an antioxidant
2. Helps in the formation of collagen, the principal protein of connective tissue.
3. Aids in iron absorption in body
4. Has role in amino acid metabolism

Deficiency Syndrome

Scurvy is the classical form of vitamin C deficiency syndrome; symptoms include microcytic anaemia, bleeding gums, loose teeth, and frequent infections, degeneration of the muscles, hysteria, and depression.

H
High speed centrifuge 202, 205, 206,
HPLC 182, 183, 184, 185, 186, 187, 188
Hydrolytic 60, 72, 73, 116, 142

I
Immunoglobulin 5, 9, 14, 17, 18, 19, 21, 23, 27, 114, 159, 177, 210
Iodine value 123
Ion exchange chromatography 29, 166, 188, 189, 192
Ionic strength 32, 108, 163, 178, 188, 189, 191, 194, 197, 225

K
Ketonic 60, 116, 117
Khoa 125, 126, 131, 132, 133

L
Lactalbumin 5, 6, 8, 14, 17, 18, 19, 20, 27, 28, 66, 114, 155, 210
Lactoferrin 21, 22, 63, 114, 210
Lactoglobulin 5, 6, 8, 14, 17, 18, 19, 20, 27, 28, 32, 66, 109, 114, 130, 221
Lactoperoxidase 7, 27, 58, 63, 71, 72, 76
Lactose 1, 2, 3, 4, 5, 6, 8, 10, 11, 12, 13, 14, 29, 30, 31, 37, 38, 39, 40, 41, 42, 43, 44, 45, 46, 47, 48, 49, 50, 76, 77, 83, 84, 85, 93, 99, 106, 110, 111, 113, 115, 127, 128, 130, 131, 132, 133, 134, 137, 139, 140, 141, 215, 216, 217, 218, 219, 220, 221, 222, 224
Lassi 126, 133, 137, 138
Lipase 5, 7, 8, 53, 59, 60, 71, 72, 73, 85, 94, 108, 116, 117, 215, 219
Lipid oxidation 56, 57, 58, 74, 85, 114
Lipoprotein 24, 73, 210, 212
Lysozyme 7, 28, 29, 71, 75, 76

M
Maillard browning 32, 58, 93, 107, 115, 132
Mastitic milk 13, 14, 15
Membrane separation process 209
MFGM 5, 6, 26, 27, 51, 52, 53, 71, 73, 77, 84, 210
Micelle structure 33, 66
Milk lipid 51, 53
Milk protein 6, 12, 13, 17, 18, 21, 27, 31, 32, 97, 98, 99, 100, 106, 107, 114, 132, 136, 137, 140, 210
Milk salt 63, 133
Milk secretion 2, 14
Mishti Doi 126, 135, 136
Mutarotation 42, 43, 44, 46

N
Non-protein nitrogenous compound 6

O
Oestrum 15
Osmotic pressure 3, 14, 37, 106, 108, 209, 216, 222, 223
Oxidative 6, 56, 60, 63, 85, 116, 215, 220

P
Page 153, 154, 155
Paneer 125, 126, 129, 130, 131
Paper chromatography 168, 170, 173, 175
Paper strip electrophoresis 162, 165
Partition chromatography 165, 166, 167, 172, 179
Phospholipid 25, 26, 55, 85, 119
Pigment 3, 6, 7, 177
Polanski value 123
Pro oxidant 71, 75
Pyridoxine 8, 227, 233, 237, 238
Pyrolysis 49

R
Rancidity 53, 56, 59, 60, 63, 72, 73, 74, 86, 97, 98, 105, 106, 113, 116, 117, 215
Rasgulla 126, 127, 128, 129
Redox potential 220, 221
Refractive index 10, 81, 82, 108, 224
Rennet 6, 14, 18, 30, 31, 32, 87, 88, 89, 90, 91, 92, 96
Rennet action 6, 89, 96
Retinol 30, 227, 228, 229
Riboflavin 4, 7, 8 15, 47, 57, 87, 133, 135, 140, 215, 233, 235, 236
RM value 123

S
Salt balance 66, 67, 68, 109
Sandesh 126, 127, 128, 129
Saponification value 123
Seaweed derivatives 100
Shrikhand 125, 126, 133, 136, 137
Specific conductance 224
Specific gravity 4, 10, 42, 216, 217
Spectrophotometry 29, 198, 200

Stabilizer 97, 100, 101
Structure 2, 18, 22, 23, 28, 29, 30, 32, 33, 34, 35, 38, 39, 41, 56, 65, 66, 76, 80, 81, 91, 99, 101, 102, 108, 120, 160, 162, 166, 183, 185, 195, 228, 229, 231, 232, 234, 235, 236, 238, 239
Surface tension 218, 219
Sweetened condensed milk 49, 105, 106, 107, 110
Sweeteners 97, 98, 99
Syneresis 90, 110, 134

T

Thermal conductivity 108, 180, 225, 226
Thermal diffusivity 226
Thiamin 140
Thin layer chromatography 168, 172, 173, 175
Tocopherol 120, 230, 231
Trace elements 7, 9, 53, 64, 132, 140, 240
Two dimensional chromatography 167, 171

U

Ultra-centrifuge 202, 206

V

Viscosity 4, 10, 42, 46, 79, 99, 100, 107, 109, 110, 141, 163, 202, 203, 207, 217, 218, 224, 225
Vitamin K 7, 52, 227, 232, 233

Y

Yogurt 139, 140, 141, 142